SuperMEN1

ADVANCES IN EXPERIMENTAL MEDICINE AND BIOLOGY

Editorial Board:
NATHAN BACK, *State University of New York at Buffalo*
IRUN R. COHEN, *The Weizmann Institute of Science*
ABEL LAJTHA, *N.S. Kline Institute for Psychiatric Research*
JOHN D. LAMBRIS, *University of Pennsylvania*
RODOLFO PAOLETTI, *University of Milan*

Recent Volumes in this Series

Volume 660
PARAOXONASES IN INFLAMMATION, INFECTION, AND TOXICOLOGY
Edited by Srinu Reddy

Volume 661
MEMBRANE RECEPTORS, CHANNELS AND TRANSPORTERS
IN PULMONARY CIRCULATION
Edited by Jason X.-J. Yuan, and Jeremy P.T. Ward

Volume 662
OXYGEN TRANSPORT TO TISSUE XXXI
Edited by Duane F. Bruley, and Eiji Takahasi

Volume 663
STRUCTURE AND FUNCTION OF THE NEURAL CELLADHESION MOLECULE
NCAM
Edited by Vladimir Berezin

Volume 664
RETINAL DEGENERATIVE DISEASES
Edited by Robert E. Anderson, Joe G. Hollyfield, and Matthew M. LaVail

Volume 665
FORKHEAD TRANSCRIPTION FACTORS
Edited by Kenneth Maiese

Volume 666
PATHOGEN-DERIVED IMMUNOMODULATORY MOLECULES
Edited by Padraic G. Fallon

Volume 667
LIPID A IN CANCER THERAPY
Edited by Jean-François Jeannin

Volume 668
SUPERMEN1
Edited by Katalin Balogh, and Attila Patocs

A Continuation Order Plan is available for this series. A continuation order will bring delivery of each new volume immediately upon publication. Volumes are billed only upon actual shipment. For further information please contact the publisher.

SuperMEN1: Pituitary, Parathyroid and Pancreas

Edited by

Katalin Balogh, MD, PhD
2nd Department of Medicine, Faculty of Medicine, Semmelweis University,
Budapest, Hungary

Attila Patocs, MD, MSc, PhD
Hungarian Academy of Sciences Molecular Medicine Research Group
and 2nd Department of Medicine, Faculty of Medicine, Semmelweis University
Budapest, Hungary

Springer Science+Business Media, LLC
Landes Bioscience

Springer Science+Business Media, LLC
Landes Bioscience

Copyright ©2009 Landes Bioscience and Springer Science+Business Media, LLC

All rights reserved.
No part of this book may be reproduced or transmitted in any form or by any means, electronic or mechanical, including photocopy, recording, or any information storage and retrieval system, without permission in writing from the publisher, with the exception of any material supplied specifically for the purpose of being entered and executed on a computer system; for exclusive use by the Purchaser of the work.

Printed in the USA.

Springer Science+Business Media, LLC, 233 Spring Street, New York, New York 10013, USA
http://www.springer.com

Please address all inquiries to the publishers:
Landes Bioscience, 1002 West Avenue, Austin, Texas 78701, USA
Phone: 512/ 637 6050; FAX: 512/ 637 6079
http://www.landesbioscience.com

The chapters in this book are available in the Madame Curie Bioscience Database.
http://www.landesbioscience.com/curie

SuperMEN1: Pituitary, Parathyroid and Pancreas, edited by Katalin Balogh and Attila Patocs. Landes Bioscience / Springer Science+Business Media, LLC dual imprint / Springer series: Advances in Experimental Medicine and Biology.

ISBN: 978-1-4419-1662-4

While the authors, editors and publisher believe that drug selection and dosage and the specifications and usage of equipment and devices, as set forth in this book, are in accord with current recommendations and practice at the time of publication, they make no warranty, expressed or implied, with respect to material described in this book. In view of the ongoing research, equipment development, changes in governmental regulations and the rapid accumulation of information relating to the biomedical sciences, the reader is urged to carefully review and evaluate the information provided herein.

Library of Congress Cataloging-in-Publication Data

SuperMEN1 : pituitary, parathyroid, and pancreas / [edited by] Katalin Balogh, Attila Patocs.
   p. ; cm. -- (Medical intelligence unit) (Advances in experimental medicine and biology ; v. 668)
 Includes bibliographical references and index.
 ISBN 978-1-4419-1662-4
 1. Endocrine glands--Tumors. 2. Pituitary gland--Tumors. 3. Parathyroid glands--Tumors. 4. Pancreas--Tumors. I. Balogh, Katalin, 1977- II. Patocs, Attila, 1973- III. Series: Medical intelligence unit (Unnumbered : 2003)
 [DNLM: 1. Multiple Endocrine Neoplasia Type 1--physiopathology. 2. Pancreatic Neoplasms--etiology. 3. Parathyroid Neoplasms--etiology. 4. Pituitary Neoplasms--etiology. 5. Tumor Suppressor Proteins--metabolism. WK 140 S959 2009]
 RC280.E55S87 2009
 616.99'44--dc22
                                           2009028832.

PREFACE

The vast expansion in research on tumorigenesis has greatly increased our understanding of tumor development in patients with inherited endocrine tumor syndromes. This book provides an up-to-date summary from clinical basics and latest follow-up guidelines to the most recent molecular findings in multiple endocrine neoplasia Type 1 syndrome. Articles have been assembled by acknowledged experts in their respective fields to provide current perspectives on the clinical and genetic backgrounds of this syndrome and to review carefully the latest discoveries concerning the possible functions and interactions of menin, the protein encoded by the *MEN1* gene, including its possible role in cell cycle regulation, hematopoiesis, and bone development. The goal of the book is also to present the most recent findings and the broadest aspects of the role of menin in tumorigenesis of the endocrine glands involved in MEN 1 syndrome (pituitary, parathyroid, endocrine pancreas and adrenal). The connection between the basic experimental and clinical points of view are highlighted through a discussion on animal models, which explores the field in both an inspiring and questioning manner with a focus on areas that remain to be clarified. Our goal was to bring together clinicians and basic researchers who represent a wide range of interests in this particular field of endocrine oncology. Presenting a comprehensive and current overview of basic experimental and clinical findings, this book can bring us closer to understanding endocrine tumorigenesis in multiple endocrine neoplasia Type 1.

Katalin Balogh, MD, PhD
Attila Patocs, MD, MSc, PhD

ABOUT THE EDITORS...

KATALIN BALOGH, MD, PhD: I graduated from Semmelweis University, Budapest as a Medical Doctor in 2001 and defended my PhD thesis on the clinical and genetic aspects of multiple endocrine neoplasia Type 1 in 2007. Working as a student at the Department of Physiology, then as a physician at the 2nd Department of Medicine, Semmelweis University, Budapest, I always had the inspiration to build bridges between patients, clinicians and researchers. Being a part of connecting different fields of medicine has always been and still remains a challenge for me. My main research interest is endocrine oncology, I am a member of the Endocrine Society and an international scholarship has afforded me the opportunity to work as a research fellow in Toronto. I hope to continue to build connections between people, continents and nations.

ABOUT THE EDITORS...

ATTILA PATOCS MD, MSc, PhD: I completed my medical and PhD training at Semmelweis University, Budapest, Hungary in 1998 and 2005, respectively. My interest has always focused on experimental laboratory work; therefore I attended and completed a biomedical engineering training at the Budapest Technical University in 2000. Between 2005-2007 I participated in a two-year postdoctoral fellowship in Dr. Charis Eng's laboratory at the Comprehensive Cancer Center of the Ohio State University and at the Genomic Medicine Institute of the Cleveland Clinic. Back in my home country from 2008 I am head of the Central Isotope Diagnostic Laboratory at the Semmelweis University and as a research associate I am working with the Molecular Medicine Research Group of the Hungarian Academy of Sciences. My research focuses on genetic and genomic characterization of endocrine tumors, on understanding of glucocorticoid signaling and on development of new, routine and molecular biological laboratory methods for diagnosis of endocrine disorders.

PARTICIPANTS

Amar Agha
Academic Department of Endocrinology
Beaumont Hospital and Royal College
of Surgeons in Ireland Medical School
Dublin
Ireland

Dheepa Balasubramanian
Department of Genetics
Case Western Reserve University
Cleveland, Ohio
USA

Katalin Balogh
2nd Department of Medicine
 Faculty of Medicine
Semmelweis University
Budapest
Hungary

Pierre-André Bédard
Department of Biology
McMaster University
Hamilton, Ontario
Canada

Lucie Canaff
Departments of Medicine Physiology
 and Human Genetics
McGill University, and Calcium
 Research Laboratory and Hormones
 and Cancer Research Unit
Royal Victoria Hospital
Montreal, Quebec
Canada

Herbert Chen
Chief of Endocrine Surgery
Section of Endocrine Surgery
Department of Surgery
The University of Wisconsin
University of Wisconsin Comprehensive
 Cancer Center
Madison, Wisconsin
USA

Colin Davenport
Academic Department of Endocrinology
Beaumont Hospital and Royal College
 of Surgeons in Ireland Medical School
Dublin
Ireland

Patrick Gaudray
Génétique, Immunothérapie, Chimie
 et Cancer (GICC)
Université François Rabelais
and
Faculté des Sciences et Techniques,
 Parc de Grandmont
Tours
France

Geoffrey N. Hendy
Calcium Research Laboratory
Royal Victoria Hospital
Montreal, Quebec
Canada

Jay L. Hess
Department of Pathology
University of Michigan Medical School
Ann Arbor, Michigan
USA

Peter Igaz
2nd Department of Medicine
Faculty of Medicine
Semmelweis University
Budapest
Hungary

Hiroshi Kaji
Division of Diabetes
Metabolism and Endocrinology
Department of Internal Medicine
Kobe University Graduate School
  of Medicine
Kobe
Japan

Terry C. Lairmore
Department of Surgery
Division of Surgical Oncology
Scott and White Memorial Hospital
  Clinic
Texas A&M University System Health
  Sciences Center College of Medicine
Temple, Texas
USA

Jean-Jacques Lebrun
Hormones and Cancer Research Unit
Department of Medicine
Royal Victoria Hospital
McGill University
Montreal, Quebec
Canada

Ivan Maillard
Department of Pathology
University of Michigan Medical School
Ann Arbor, Michigan
USA

Bart M. Maslikowski
Department of Biology
McMaster University
Hamilton, Ontario
Canada

Maria Papaconstantinou
Department of Biology
McMaster University
Hamilton, Ontario
Canada

Attila Patocs
Hungarian Academy of Sciences
  Molecular Medicine Research Group
  and 2nd Department of Medicine
Faculty of Medicine
Semmelweis University
Budapest
Hungary

Alicia N. Pepper
Department of Biology
McMaster University
Hamilton, Ontario
Canada

Karoly Racz
2nd Department of Medicine
Semmelweis University
Budapest
Hungary

Peter C. Scacheri
Department of Genetics
Case Western Reserve University
Cleveland, Ohio
USA

Günther Weber
Génétique, Immunothérapie, Chimie
  et Cancer (GICC)
Université François Rabelais
and
CNRS UMR 6239
Faculté des Sciences et Techniques
Tours
France

CONTENTS

1. MEN1 CLINICAL BACKGROUND 1
Peter Igaz

Abstract 1
Introduction 1
Clinical Features 2
Diagnosis of MEN1 6
MEN1 Variants and Phenocopy 6
Diagnostics of MEN1-Related Tumors 8
Screening for MEN1 Manifestations 9
Indications for *MEN1* Germline Mutation Screening 9
Therapy 11
Surveillance 12
Comments and Conclusion 12

2. GENETIC BACKGROUND OF MEN1: FROM GENETIC HOMOGENEITY TO FUNCTIONAL DIVERSITY 17
Patrick Gaudray and Günther Weber

Abstract 17
Introduction: The History of a Rare Endocrine Genetic Disease 17
On the Nature of the *MEN1* Gene 18
On the Regulation of the *MEN1* Gene 20
What Do We Learn from the Hereditary Mutations of the *MEN1* Gene? ... 21
Importance of MEN1 in Endocrine Tumorigenesis 21
Is MEN1 a Genome Instability Syndrome? 22
Conclusion 23

3. MENIN: THE PROTEIN BEHIND THE MEN1 SYNDROME 27
Maria Papaconstantinou, Bart M. Maslikowski, Alicia N. Pepper
and Pierre-André Bédard

Abstract 27
Introduction 27
Menin Is a Nuclear Protein—Role of the C-Terminal Region 27

xiii

Leucine-Rich Domains in Menin 29
GTPase Signature Motif... 30
Post-Translational Modification in Response to DNA Damage 30
Conservation of Menin Structure, Protein Interactions and Function......... 30
Conclusion 33

4. CELLULAR FUNCTIONS OF MENIN... 37
Geoffrey N. Hendy, Hiroshi Kaji and Lucie Canaff

Abstract.. 37
Introduction 37
Cell Cycle 37
Cell Cycle Checkpoints and DNA Repair 43
Chromatin Remodeling 45
Conclusion 47

5. THE ROLE OF MENIN IN HEMATOPOIESIS................................ 51
Ivan Maillard and Jay L. Hess

Abstract.. 51
Introduction 51
Role of Menin in Hematopoiesis 53
Role of Menin in Leukemogenesis 54
Conclusion 55

6. ROLE OF MENIN IN BONE DEVELOPMENT 59
Hiroshi Kaji, Lucie Canaff and Geoffrey N. Hendy

Abstract.. 59
Introduction 59
Menin and TGF-β Signaling 59
Menin and AP-1 Signaling.. 60
BMP, TGF-β and AP-1 Signaling in the Osteoblast 60
Role of Menin in Early Stage Osteoblast Differentiation 62
Menin and TGF-β Pathway in Osteoblast Differentiation 62
Menin and JunD in the Osteoblast 64
Conclusion 65

7. ACTIVIN, TGF-β AND MENIN IN PITUITARY TUMORIGENESIS 69
Jean-Jacques Lebrun

Abstract.. 69
Introduction 69
The Activin/TGF-β Superfamily.. 71
Activin/TGF-β in the Pituitary 72
Activin Inhibits Prolactin Gene Expression and Signalling............................ 72
Loss of Menin Inhibits TGF-β Induced Transcriptional Activity.................. 73
Menin Interacts with Smad Proteins... 73
Smads and Menin Are Required for Activin-Mediated Cell Growth Inhibition
    and Repression of Prolactin Gene Expression... 74
Conclusion 75

8. THE ROLE OF MENIN IN PARATHYROID TUMORIGENESIS 79
Colin Davenport and Amar Agha

Abstract 79
Introduction 79
The *MEN1* Gene 79
MEN1 Related Mutations and Menin Expression in Hereditary and Sporadic
  Hyperparathyroidism (Genotype-Phenotype Correlation) 82
TGF-β/Smad3 Signalling 83
Menin and TGF-β Signalling 83
Other Forms of Parathyroid Tumorigenesis 85
Conclusion 85

9. ROLE OF MENIN IN NEUROENDOCRINE TUMORIGENESIS 87
Terry C. Lairmore and Herbert Chen

Abstract 87
Introduction 87
Global Gene Expression in Normal Islet Cells versus MEN 1-Associated
  Neuroendocrine Tumors 88
Signaling Pathways in Neuroendocrine Tumors 90
Treatment of Neuroendocrine Tumors Based on Molecular Genetic Diagnosis 92

10. ADRENAL TUMORS IN MEN1 SYNDROME AND THE ROLE
OF MENIN IN ADRENAL TUMORIGENESIS 97
Attila Patocs, Katalin Balogh and Karoly Racz

Introduction 97
Genetics of Adrenal Tumors 97
Hereditary Syndromes with Adrenal Involvement 97
Somatic Genomics of Sporadic Adrenal Tumors 98
MEN1-Associated Adrenal Tumors 99
Diagnosis, Therapy and Follow-Up of Adrenal Tumors 100
MEN1 Gene Mutation Screening in Patients with Adrenal Tumors:
  To Screen or Not? 101
Comments and Conclusion 102

11. FUNCTIONAL STUDIES OF MENIN THROUGH GENETIC
MANIPULATION OF THE *MEN1* HOMOLOG IN MICE 105
Dheepa Balasubramanian and Peter C. Scacheri

Abstract 105
Introduction 105
Conventional *Men1* Mouse Models 106
Conditional *Men1* Mouse Mutants 108
Menin Overexpression 110
Crossbreeding Studies 111
Conclusion 112

INDEX 117

ACKNOWLEDGEMENTS

Finally, the editors would like to acknowledge all the authors and colleagues who have contributed in various ways to this book, and we would like to express our appreciation to the staff at Landes Bioscience for their cooperation.

Chapter 1

MEN1 Clinical Background

Peter Igaz*

Abstract

Multiple endocrine neoplasia Type 1 (MEN1) is a rare hereditary tumor syndrome predisposing to tumor development in several endocrine organs. Its major manifestations include hyperparathyroidism, tumors of endocrine pancreas and pituitary. Beside these three, several other endocrine (adrenocortical, foregut carcinoid) and nonendocrine (lipoma, angiofibroma, collagenoma, ependymoma, meningioma) tumors have been described to be associated with this syndrome. Both familial and sporadic forms of the disease are known. The diagnosis of MEN1 can be established if two of the three major manifestations are found in the same patient, whereas the diagnosis of familial MEN1 requires one MEN1 patient and a first degree relative with at least one MEN1 manifestation. MEN1 is transmitted as an autosomal dominant trait with high penetrance, approaching 95-100% by the age of 60. Both benign (parathyroid, anterior pituitary) and malignant (gastrinoma, glucagonoma) lesions may develop in MEN1 patients. Regular surveillance of *MEN1* gene mutation carriers is necessary to reveal disease manifestations. Several diagnostic modalities can be used to screen for and to examine MEN1-related tumors. The therapy of MEN1-associated tumors requires specific approach in some cases, as multiple tumors and recurrence is frequently observed.

Introduction

Multiple endocrine neoplasia syndrome Type 1 (MEN1, OMIM 131100) has been first described by Wermer in 1954 as an association of tumors of several endocrine organs.[1] The main organs affected include the parathyroids, the endocrine pancreas and pituitary. Apart from these three major organs, many other, including some nonendocrine tumors may develop in the affected patients.[2-7]

MEN1 is a hereditary disorder with autosomal dominant transmission, therefore a child of a MEN1 affected parent has 50% chance of inheriting the disease. The development of multiple endocrine tumors in the same patient is of great significance also from a scientific point of view, that may indicate common points in the tumorigenesis of different endocrine organs. The genetic background of MEN1 was elucidated in 1997 by the identification of a putative tumor suppressor gene (*MEN1* gene, on chromosome 11q13), whose inactivating mutations have already been found in the majority of MEN1 patients.[8] Unfortunately, no mutation hotspots (i.e., gene regions, where mutations are frequently found) have been revealed in the *MEN1* gene and efforts at establishing genotype-phenotype correlations have also been largely unsuccessful to date. The *MEN1* gene codes for a protein termed menin that is involved in numerous molecular biological pathways which are discussed in other chapters in detail.[9] Besides its inherited form, MEN1 is also observed in a sporadic setting, mostly due to de novo mutations. The likelihood of finding an *MEN1* gene mutation is about 70-80% in familial cases, whereas only 45-50% of sporadic cases were found to harbor

*Peter Igaz—2nd Department of Medicine, Faculty of Medicine, Semmelweis University, 1088 Budapest, Szentkirályi u. 46, Hungary. Email: igapet@bel2.sote.hu

SuperMEN1: Pituitary, Parathyroid and Pancreas, edited by Katalin Balogh and Attila Patocs. ©2009 Landes Bioscience and Springer Science+Business Media.

Table 1. Comparison of MEN1 and MEN2 syndromes

	MEN1	MEN2
Function of responsible gene	Tumor suppressor	Protooncogene
Type of mutations	Inactivating	Gain of function
Inheritance	AD	AD
Mutation hotspots	No	Yes
Penetrance	High	High
Genotype-Phenotype correlations	No	Yes
Likelihood of finding germline mutations in familial cases	70-80%	>90%
Indication of prophylactic surgery	No*	Yes (MTC)

*As MEN1-related tumors arise in indispensable organs (pituitary, endocrine pancreas) or the disease can be efficiently diagnosed and cured (HPT), prophylactic surgery is not indicated, except for prophylactic thymectomy performed during parathyroid surgery to prevent the development of thymic carcinoids.

MEN1 mutations. The chance of finding an *MEN1* gene mutation in an individual increases with the number of MEN1-related tumors present in the patient and with a positive family history.[10]

There are several other monogenic tumor syndromes that include tumors of endocrine organs, as well. It is interesting to note that almost all such syndromes show an autosomal dominant inheritance pattern.

Manifestations of the other major multiple endocrine neoplasia syndrome (multiple endocrine neoplasia type 2, MEN2) include medullary thyroid cancer (MTC), pheochromocytoma and hyperparathyroidism (HPT). MEN2 is caused by gain-of-function mutations of the RET (Rearranged during Transfection) protooncogene. MEN2 has three forms: (i) MEN2A is an association of MTC, pheochromocytoma and HPT; (ii) in MEN2B a very aggressive MTC variant may associate with pheochromocytoma without HPT and patients show characteristic marfanoid appearance with mucocutaneous neurinomas; (iii) in familial MTC (FMTC) MTC is the only manifestation. Significant genotype-phenotype correlations have been established in MEN2 syndrome, i.e., it is possible to predict the expected phenotype in individuals with certain mutations. The likelihood of finding germline mutations in MEN2 patients is higher than in MEN1. In individuals carrying RET mutations, prophylactic thyroidectomy is indicated for the prevention of MTC development.[4,6] The major differences between MEN1 and MEN2 syndromes are summarized in Table 1.

Table 2 summarizes the major manifestations of some of the tumor syndromes with endocrine relevance.

In the following, the clinical features of MEN1 syndrome and questions related to its diagnostics, therapy and follow-up will be discussed.

Clinical Features

MEN1 is a rare syndrome with an overall prevalence of approximately 1:30.000.[4,6] Its penetrance is high (i.e., the likelihood of overt disease in an individual carrying a *MEN1* gene mutation), reaching 95% by the age of 55 years. MEN1 is rare in children and young adolescents, the typical age of its diagnosis is around 20-30 years.[5]

The most common and most characteristic manifestation of MEN1 is hyperparathyroidism. In contrast with its sporadic counterpart, MEN1-associated HPT is mostly caused by the asymmetric hyperplasia of all parathyroid glands or multiple tumors. Ectopic, mainly mediastinal location is also common. HPT in MEN1 manifests itself decades earlier than sporadic forms, often at the age of 20-25 years, therefore HPT in a patient under 30 years should raise the suspicion of MEN1.

Whereas sporadic HPT is more frequent in females (3:1 male: female ratio), gender imbalance is not observed in MEN1-related HPT. Symptoms do not differ between MEN1-related and sporadic HPT that include skeletal (osteoporosis, fractures), kidney (polyuria, polydipsia, nephrolithiasis, nephrocalcinosis), gastrointestinal (constipation, nausea, stomach and duodenal ulcers, acute and chronic pancreatitis), cardiovascular (accelerated arteriosclerosis, ischaemic heart disease) and central nervous system (confusion, dementia, psychosis) manifestations,[11] but recurrence is more common with MEN1-related HPT.[4-6] HPT is observed in nearly all MEN1 patients by the age of 60 years. MEN1-related parathyroid tumors are almost never malignant.

Tumors of the endocrine pancreas comprise the second most common type of neoplasms in MEN1. Their prevalence varies between 30 and 80% in different studies.[4,12,13]

Gastrinomas are the most frequent, with prevalence reaching 40%. Gastrin oversecretion results in the classic Zollinger-Ellison syndrome characterized by recurrent gastric and duodenal ulcers. Diarrhoea is also a characteristic feature. Gastrinomas are mostly malignant, often metastatising to the liver. MEN1-related gastrinoma manifests itself about a decade earlier than its sporadic counterpart. In contrast with sporadic HPT that is the third most common endocrine disease in the adult population, gastrinoma is rare and a significant proportion of sporadic gastrinomas (≈ 25%) form part of the MEN1 syndrome. MEN1-related gastrinomas are mostly multifocal with more than 50% located in the duodenum. Liver metastases are found in 23% of patients.[14] The concomitant HPT can aggravate symptoms of hypergastrinemia. 50% of gastrinomas have already metastastised before diagnosis. Gastrinomas are major mortality factors in MEN1, responsible for about 30% of deaths in MEN1.[14]

Insulinomas occur in 10% of MEN1 patients. They are mostly benign, however, their metabolic consequences can be life-threatening. Fasting hypoglycaemia, symptoms of neuroglycopenia and significant weight gain are its cardinal features.[4,12]

Other endocrine pancreatic tumors, i.e., VIPoma, glucagonoma, somatostatinoma and GRF-oma, may occur as infrequent manifestations of MEN1. VIPoma (Werner-Morrison syndrome) is also known as pancreatic cholera, as the oversecretion of vasoactive intestinal peptide (VIP) results in excessive, watery diarrhoea and hypokalemia (WDHA syndrome: watery diarrhoea, hypokalemia, achlorhydria).[15] Glucagonoma is almost always malignant with impaired glucose tolerance or diabetes mellitus and a characteristic skin lesion (migrating necrolytic erythema) is often observed.[16] The symptoms of somatostatinoma include hyperglycemia, diarrhoea, steatorrhoea, cholelithiasis.[17] Growth hormone releasing factor (GRF) secreting tumors (GRF-oma) may lead to the development of acromegaly. Its diagnosis is often established following an unsuccessful pituitary operation.[13]

A significant portion (20-30%) of endocrine pancreas tumors are nonfunctioning, or secrete pancreas polypeptide (PP). PP-hypersecretion, however, has not yet been associated with clinical symptoms.[12,13]

Anterior pituitary tumors constitute the third most important MEN1 syndrome manifestation. They are observed in 15-50% (15-90% by others) of MEN1 patients. The majority (~60%) of pituitary tumors secrete prolactin (PRL, prolactinoma). Growth hormone (GH)-secreting tumors (~25%), resulting in acromegaly are also observed. Some tumors secrete both PRL and GH. Adrenocorticotropin (ACTH) secreting tumors are rare in MEN1 and lead to the development of hypercortisolism (Cushing's disease). Besides hormone secretion, pituitary tumors may cause symptoms due to mass effects (visual field anomalies, cranial nerve palsies, increased intracranial pressure) and hormone deficiency (hypopituitarism) may also develop. Two-thirds of tumors are microadenomas, with a diameter less than 1 cm.[4-6] MEN1-related pituitary adenomas are usually larger and behave more aggressively (faster growth, invasiveness) than sporadic pituitary tumors.[18]

Adrenal tumors are observed in 20-40% of MEN1 patients. These originate mainly from the adrenal cortex. Their vast majority is hormonally inactive, a small proportion (<10%) can be hormone secreting leading to hypercortisolism (Cushing's syndrome). Primary aldosteronism has also been occasionally reported. The adrenocortical tumors are mostly benign, adrenocortical cancer (ACC) has been described only in a few cases. Pheochromocytomas also occur in MEN1 (<1%), but they cannot be considered as major manifestations.[2-4,19]

Table 2. Major features of tumor syndromes with endocrine involvement

Tumor Syndrome	Transmission	Responsible Gene	Gene Function	Major Endocrine Organs Affected and Endocrine Manifestations	Major Nonendocrine Manifestations
Multiple endocrine neoplasia Type 1 (MEN1)	AD	*MEN1* gene	Tumor suppressor	Parathyroid, endocrine pancreas, pituitary, adrenal	Collagenoma, facial angiofibroma, ependymoma
Multiple endocrine neoplasia Type 2 (MEN2) MEN2A MEN2B FMTC	AD	RET	Protooncogene	Medullary thyroid cancer, pheochromocytoma, parathyroid (HPT)	Mucocutaneous neurinomas in MEN2B
Von Hippel Lindau disease (VHL)	AD	VHL	Tumor suppressor	Pheochromocytoma, tumors of the endocrine pancreas (rare)	Kidney tumor, retinal and cerebellar haemangioblastomas, exocrine pancreas tumors and cysts, epididymis cysts
Neurofibromatosis Type 1 (NF1)	AD	NF-1	Tumor suppressor	Pheochromocytoma, endocrine pancreas tumors (somatostatinoma)	Neurofibromas, café-au-lait spots
Hereditary paragaglioma (PGL) syndromes	AD	SDHC, SDHD	Tumor suppressor	Pheochromocytoma, paraganglioma (extraadrenal pheochromocytoma)	
Carney complex (CC)	AD	PRKAR1A and others	Tumor suppressor	PPNAD (Cushing's syndrome), pituitary (acromegaly), thyroid tumor, Sertoli cell tumor	Cardiac and breast myxomas, lentiginosis
McCune-Albright syndrome	Non-transmitted*	GNAS	Tumor suppressor	Pituitary (acromegaly, prolactinoma), precocious puberty, adrenal tumor	Polyostotic fibrous dysplasia

continued on next page

Table 2. Continued

Tumor Syndrome	Transmission	Responsible Gene	Gene Function	Major Endocrine Organs Affected and Endocrine Manifestations	Major NonEndocrine Manifestations
Hyperparathyroidism-jaw tumor syndrome	AD	parafibromin	Tumor suppressor	HPT	Fibrous jaw tumor, Wilms-tumor, kidney tumor, polycystic kidneys
Li-Fraumeni syndrome	AD	p53	Tumor suppressor	Adrenocortical cancer	Soft tissue sarcoma, breast cancer, leukemia, brain tumor
Cowden syndrome	AD	PTEN	Tumor suppressor	Thyroid cancer (mostly follicular)	Multiple hamartomas, breast cancer, endometrial cancer
Tuberous sclerosis	AD	TSC1, TSC2	Tumor suppressor	Tumors of endocrine pancreas (rare)	Multiple hamartomas (brain, skin, retina, kidney, heart, lung)

AD: autosomal dominant; FMTC: familial medullary thyroid carcinoma; PPNAD: primary pigmented nodular adrenal disease; SDH: succinate dehydrogenase.
*Germline mutations of the GNAS gene are thought to be incompatible with life, GNAS mutations in McCune-Albright syndrome arise during embryonic development.

Foregut carcinoid tumors represent a significant, but not as frequent manifestations of MEN1 as the preceding features. Thymic carcinoids are more frequent in male patients, often malignant and large at diagnosis.[20] Underlying MEN1 may be found in about 25% of sporadic thymic carcinoids. Other carcinoids (bronchial, gastric, duodenal) are also found. Both gastric and bronchial MEN1-related carcinoids can be more aggressive than their sporadic counterparts. Whereas more than 50% of sporadic carcinoids are hormone-secreting, MEN1-related carcinoid tumors are mostly hormonally inactive, very rare ACTH, CRH (corticotropin releasing hormone) or GRF secreting tumors may be associated with hypercortisolism and acromegaly, respectively.[4,21]

Thyroid tumors have been reported in over 25% of MEN1 patients, but as these are very frequent in the population, they are not considered to form an integral part of MEN1 syndrome.[22]

Apart from endocrine tumors, MEN1 syndrome encompasses nonendocrine manifestations, as well. Very frequently observed dermatological manifestations include collagenomas, lipomas and facial angiofibromas. These "minor" lesions can be very important for diagnosis. The presence of one or more collagenomas and 3 or more angiofibromas in MEN1 patients with gastrinomas was found to have 75% sensitivity and 95% specificity for MEN1.[23] Central nervous system tumors i.e., ependymomas and meningiomas have also been described.[24] The tumor manifestations of MEN1 are summarized in Table 3. Some endocrine syndromes may develop via multiple pathogenetic mechanisms, described in Table 4.

Diagnosis of MEN1

The diagnosis of MEN1 can be established if two of the three major features (HPT, endocrine pancreas, pituitary) are found in the same individual either in a simultaneous or a metachronous fashion. The diagnosis of familial MEN1 is based on a MEN1 patient having a first degree relative with at least one of the major manifestations.[4,5]

MEN1 Variants and Phenocopy

As MEN1 includes many manifestations, the identification of some variants with peculiar characteristics is not surprising. In fact, MEN1 is one of the most variable endocrine tumor syndromes.[25]

The MEN1 Burin variant was identified in four large kindreds in Newfoundland (Canada) harboring the same mutation, that raises the possibility of a founder effect. Prolactinoma, carcinoid and late onset parathyroid tumors are the most characteristic features in these patients, endocrine pancreas tumors are rarely observed.[26] Apart from the Burin variant, other families, harboring other mutations with frequent prolactinoma and rare gastrinoma manifestations were identified.[27]

Hereditary HPT can be part of MEN1, MEN2, familial hypocalciuric hypercalcemia (FHH) and hyperparathyroidism-jaw tumor syndromes.[28] Apart from these, the entity of familial isolated hyperparathyroidism (FIHP) has also been described in several kindreds.[29,30] Some studies revealed *MEN1* gene mutations in about 20-35% of FIHP kindreds,[10,31] but other studies failed to confirm these findings.[32]

A low penetrance MEN1 family with acromegaly and HPT was also reported.[33] Familial isolated pituitary adenoma (FIPA) syndrome has not been confirmed to be a MEN1 variant.[34]

The phenomenon of phenocopy is known to occur in MEN1. Phenocopy describes the situation if the clinical diagnosis of a genetically determined disease is established in an individual without the specific genetic alteration, i.e., a sporadic form of the disease is mistakenly judged as genetically determined. MEN1 phenocopy occurs both in sporadic and familial settings. As sporadic forms of mild HPT and asymptomatic pituitary tumors are frequent, MEN1 family members suffering from these but lacking specific *MEN1* mutations can be incorrectly judged as being MEN1 patients. Up to 10% of MEN1 patients diagnosed by clinical, biochemical and radiological screening may have phenocopy.[35] Theoretically, MEN1 phenocopy may also be caused by other familial tumor syndromes with low penetrance, or—similar to McCune-Albright's syndrome—due to somatic mutations during early embryonic stages.[36] In most cases, phenocopy can be ruled out by genetic analysis. Patients with MEN1 phenocopy do not need surveillance for MEN1 manifestations.

Table 3. Manifestations of MEN1 syndrome

Endocrine Manifestations		Nonendocrine Manifestations	
Feature	**Percentage**	**Feature**	**Percentage**
HPT	90-100[4,6]	Facial angiofibroma	85-90[4]
Endocrine Pancreas Tumors	30-80[22]	Collagenoma	70[4]
Gastrinoma	40[4]	Lipoma	20-30[4,22]
Insulinoma	10[4]	Leiomyoma (esophagus, lung, rectum, uterus)	10[55]
Nonfunctioning	20[4]	Ependymoma	1[4]
Glucagonoma, VIPoma, GRFoma, somatostatinoma	2[4]	Meningioma	5-8[4,24]
Anterior Pituitary Tumors	15-90[7]		
Prolactinoma	20[4]		
Nonfunctioning	5[4]		
GH-secreting	5[4]		
GH + PRL secrering	5[4]		
ACTH-secreting	2[4]		
TSH-secreting	<1[4]		
Adrenal Tumors			
Adrenocortical adenoma	5-40[4,7,22]		
ACC	<1[21]		
Pheochromocytoma	<1[4]		
Foregut Carcinoids			
Thymic	2[4]		
Gastric	8-10[4]		
Bronchial	2[4]		

Table 4. Endocrine syndromes with complex etiology in MEN1

Endocrine Syndrome	Etiology
Acromegaly	i. GH-secreting pituitary tumors
	ii. GRF-secreting endocrine pancreatic tumors (rare)
	iii. GRF-secreting bronchial carcinoid (rare)
Hypercortisolism (Cushing's disease or syndrome)	i. ACTH-secreting pituitary tumors
	ii. Cortisol-secreting adrenocortical tumor (rare)
	iii. CRH or ACTH secreting foregut carcinoid (rare)

Diagnostics of MEN1-Related Tumors

HPT

The laboratory diagnosis of HPT is in general easy and relies on elevated serum calcium (hypercalcemia) and serum parathormone levels.[11] As all four glands tend to be affected in MEN1 that warrants intraoperative surgical examination of all, preoperative imaging is of lesser importance than in case of sporadic parathyroid tumors.

Imaging techniques are indispensable, if reoperation is required. Ultrasonography, ^{99m}Tc-sestamibi (methoxyisobutylisonitrile) parathyroid scintigraphy, computed tomography (CT) or magnetic resonance imaging (MRI) can be used. Among these, the combination of scintigraphy and ultrasonography is the most sensitive (>90%) for locating neck lesions, the identification of mediastinal tumors is difficult, scintigraphy and MRI may be helpful.[37]

Pancreatic Islet Tumors

Laboratory Diagnosis

The laboratory diagnosis of gastrinoma relies on high serum gastrin levels. Gastric pH measurement and dynamic tests (secretin and calcium infusion test) have diminishing clinical relevance, but can be well exploited if the diagnosis is umbiguous. As drugs inhibiting gastric acid secretion (H2-antagonists, proton pump inhibitors (PPI)) augment serum gastrin levels, these should be discontinued before its measurement (two days for H2 antagonists, one week for PPI).[38]

For the diagnosis of insulinoma, the simultaneous elevation of serum insulin and C-peptide concentrations with hypoglycemia (se Glu<2.2 mmol/l) are mandatory. The 72-hours fasting test is most widely used for the provocation of hypoglycemia.[39] Elevated serum proinsulin concentration can be characteristic for insulinoma.[40]

For the laboratory diagnosis of the other hormone-secreting pancreatic tumors, the measurement of the specific hormone (VIP, GRF, somatostatin) is the most appropriate.[41]

Most tumors of the endocrine pancreas secrete chromogranin proteins. Chromogranins are stored in the dense granules of neuroendocrine cells and released along with the mediators. Chromogranin A (CgA) is a sensitive marker for many endocrine pancreas tumors, including gastrinoma and VIPoma, whereas chromogranin B (CgB) is more characteristic for insulinomas.[42-44]

Hormonally inactive tumors may also secrete CgA that can be well exploited for their laboratory diagnosis. Pancreatic polypeptide (PP) may also be secreted, but its sensitivity and specificity is lower than that for CgA.[41] CgA, however, can be elevated in other tumors too, including carcinoid tumors, pheochromocytomas, other neoplasms (hepatocellular, lung, prostate cancer). Essential hypertension, chronic renal insufficiency, steroid therapy and gastric acid suppression (H2-antagonist and PPI therapy) may also induce serum CgA elevation that limit its specificity.[45]

Imaging

As pancreatic islet tumors are mostly small, their localization is difficult. Besides classic imaging techniques (ultrasonography, CT, MRI), specific diagnostic modalities can also be exploited. Among these, somatostatin receptor scintigraphy can be very helpful, as the majority of these tumors express receptors for somatostatin (SSR1-5) enabling the use of radioactively labelled somatostatin analogues (^{111}In-diethylenetriamine pentaacetic acid octreotide) for diagnostic purposes. Endoscopic ultrasound is very sensitive (>90%) for the localization of duodenal or pancreatic tumors. Angiography is less used nowadays, but selective arterial stimulation can be efficient for the diagnosis of insulinoma: selective injection of calcium in the catheterised pancreatic supplying arteries leads to insulin secretion that can be measured in the blood taken from hepatic veins.[46-49]

Pituitary Tumors

Detailed endocrinological examinations are needed for the analysis of hormone-secretion or hormone deficiency states. Serum prolactin, thyroid stimulating hormone (TSH), ACTH, cortisol, gonadotropins (luteinizing hormone (LH), follicle stimulating hormone (FSH)), insulin like growth factor (IGF-1) should be routinely examined. IGF-1 can be sufficient to screen for developing acromegaly, however, the correct diagnosis of acromegaly relies on oral glucose tolerance test (OGTT) with serial growth hormone (GH) measurements.

MRI is the gold standard imaging technique for pituitary tumors, enabling the diagnosis of even small microadenomas.[50]

Adrenal Tumors

Adrenal tumors are mostly discovered by routine imaging techniques, i.e., ultrasonography or CT. Although these are mostly hormonally inactive, hormone secretion should be excluded. Urinary cortisol, midnight-morning cortisol rhythm and the low-dose dexamethasone suppression test can be applied for the diagnosis of Cushing's syndrome, renin/aldosterone ratio is used for examining primary aldosteronism. Adrenal androgens should also be determined.

Pheochromocytoma does not constitute a major MEN1 manifestation, on clinical suspicion, urinary catecholamine metabolite measurements (metanephrine, normetanephrine, homovanillic acid, vanillylmandelic acid) or serum CgA determination should be performed. ^{131}I-MIBG (meta-iodo-benzyl guanidine, a catecholamine precursor) scintigraphy is a highly specific imaging technique for pheochromocytoma diagnosis.[51]

Carcinoid Tumors

Most MEN1-related carcinoid tumors are hormonally inactive. However, serum CgA can be elevated even in hormonally inactive tumors and urinary 5 HIAA (5-hydroxy-indol acetic acid, a serotonin metabolite) should also be determined. As thymic carcinoid tumors are often indolent and large at diagnosis, chest CT should be regularly performed in MEN1 patients to diagnose this potentially fatal manifestation. Thymic carcinoids may be locally invasive and indistinguishable from a thymoma.[20] SSR scintigraphy is useful for the diagnosis of carcinoids.

Nonendocrine Tumors

Dermatological tumors are easily diagnosed on physical examination. As no laboratory parameters are known for smooth muscle and central nervous system tumors, these can only be diagnosed by imaging techniques.

Screening for MEN1 Manifestations

MEN1 patients should be screened regularly to reveal disease manifestations. Periodical clinical screening of *MEN1* gene mutation carriers is effective to detect MEN1-related tumors, therefore it may reduce morbidity and mortality.[52]

According to the recommendations of the MEN1 consensus conference,[4] annual biochemical and periodical (3-5 years) imaging screenings should be performed. HPT screening should begin at the age of 8 years by annual serum calcium and PTH measurements. Screening for gastrinoma should be started at age 20, annual serum gastrin and periodical (every 3-5 years) abdominal CT or MRI can be performed. Insulinoma may appear in earlier ages, therefore its screening should already begin at the age of 5 by fasting glucose, insulin and C-peptide determinations. Annual CgA, glucagon and PP determinations may be used for screening other enteropancreatic manifestations that should begin at age 20. SSR scintigraphy, CT or MRI can be used as imaging techniques. Screening for pituitary tumors should also begin at age 5, serum prolactin and IGF-1 should be determined as the most frequently oversecreted hormones, pituitary MRI is the most efficient imaging modality. Foregut carcinoids can be screened by serum CgA, starting at the age of 20. Chest CT is proposed to search for thymic carcinoids. There is no consensus regarding adrenocortical tumors, no serum marker can be proposed, periodical abdominal CT can be suggested (Table 5).

Indications for *MEN1* Germline Mutation Screening

In an index case, *MEN1* mutation screening is always indicated if the clinical criteria of MEN1 are met either in a familial or a sporadic setting.

Table 5. Screening for MEN1 manifestations in MEN1 mutation carriers

Age to Begin at (Years)	Manifestation	Biochemical Test (Annual)	Imaging (Every 3-5 Years)
5	Insulinoma	Fasting glucose, insulin, C-peptide	-
5	Pituitary	Serum prolactin, IGF-1	MRI
8	HPT	Serum calcium, PTH	-
20	Gastrinoma	Serum gastrin	Abdominal CT or MRI
20	Other enteropancreatic	CgA, glucagon, proinsulin, PP	SSTR scintigraphy, CT, MRI
20	Foregut carcinoid	CgA	CT, SSTR scintigraphy
?	Adrenal tumor	?	CT

Asymptomatic individuals in MEN1 families with known germline mutations should be examined. The presence of a *MEN1* mutation warrants the need for regular clinical follow-up, whereas absence of a known mutation in a familial setting renders further clinical screening unnecessary.

MEN1 mutation analysis can be performed in cases where the diagnosis of MEN1 cannot be established on the clinical criteria, but the case is suspicious or atypical of MEN1.[4] The prevalence of underlying MEN1 in case of sporadic endocrine tumors is summarized in Table 6. Genetic analysis for MEN1 should be performed in young (<30 y) HPT patients, or if recurrent disease and multiple tumors are found. As the percentage of underlying MEN1 is high with sporadic gastrinomas and thymic carcinoids, its routine genetic analysis can be proposed. *MEN1* genetic analysis is not indicated in sporadic pituitary and adrenal tumors. Unfamiliar clinical setting (e.g., multiple islet tumors at any age) also warrants genetic examination.

Considering the molecular biological methods to be used, only the large scale automated bidirectional sequencing of the whole *MEN1* gene can be proposed for the analysis of *MEN1* mutational status in a sporadic MEN1 patient. Direct sequencing, however, is very expensive and laborious, therefore other methods were developed for *MEN1* mutation screening. These methods can be used as "prescreening" to facilitate sequencing by identifying the exons that carry genetic alterations and also for screening members of MEN1 families with known mutations. Pre-screening methods are summarised in Table 7. For family screening, allele-specific

Table 6. Prevalence of underlying MEN1 syndrome in sporadic tumors

Tumor	Prevalence of Underlying MEN1 (Reference)
Parathyroid	1%[4,5]
Gastrinoma	25%[3]
Insulinoma	10%[6]
Pituitary tumor	<3%[3]
Adrenal tumor	<1%[62]
Thymic carcinoid	25%[21]
Other foregut carcinoid	<4%[63]

Table 7. Comparison of genetic prescreening methods for MEN1

Method	Abbreviation	Sensitivity	Specificity	Cost
Single-strand conformation polymorphism	SSCP	80-90%	ND	Moderate
Fluorescent SSCP	F-SSCP	~95%	~97%	Equipment expensive, measurement cheap
Denaturing gradient gel electrophoresis	DGGE	~90%	ND	Moderate
Temperature gradient gel electrophoresis	TGGE	~90%	ND	Moderate
Conformation sensitive gel electrophoresis	CSGE	~96%	ND	Moderate
Denaturing high-performance liquid chromatography	DHPLC	~100%	~100%	Equipment expensive, high throughput measurement cheap

ND: no data available in MEN1 screening.

oligonucleotide hybridization (ASO), allele-specific amplification (ASA) and restriction fragment length polymorphism (RFLP) can be applied.[53]

Therapy

Hyperparathyroidism

Either subtotal or total surgical removal of all parathyroid glands is required in MEN1 patients, with the heterotopic autotransplantation of parathyroid tissue under the skin of a forearm to prevent hypoparathyroidism. Recurrence is common. The main goals of surgery include: (i) maintaining eucalcemia for the longest time possible, (ii) avoiding iatrogenic hypocalcemia and other major complications, (iii) facilitating future surgery for recurrent disease.[54,55] Some authors stand for the necessity of a simultaneously performed transcervical thymectomy to prevent the development of thymic carcinoids.[56] This can be considered as the only possibility of prophylactic surgery in MEN1.

A recently developed drug targeting the calcium sensing receptor, the calcimimetic cinacalcet could be an efficient drug alternative in HPT therapy,[57] but no experience is available to date in MEN1 patients.

Endocrine Pancreas Tumors

Surgical removal is the only curative therapy, but it is often not feasible if the primary tumor cannot be localized or metastases are present.

Other treatment modalities exist that may be applied to alleviate symptoms and inhibit or reduce tumor growth.

The majority of endocrine pancreas tumors express several types of somatostatin receptors. Somatostatin (SS) inhibits the hormone secretion and growth of neuroendocrine cells and tumors. Due to its short half-life, however, native somatostatin is not suitable for treatment. Somatostatin analogues (octreotide, lanreotide) can be efficiently applied for treating almost all forms of endocrine pancreas tumors. Long-acting preparations have also been developed enabling the administration of somatostatin analogues every two or four weeks.[41,58] Interferons are also effective both in controlling hormone symptoms and also in inhibiting tumor growth, but several side-effects limit their wide-spread application.[41]

Besides SS analogues and interferons, other drugs are also effective options for certain tumors. Gastric acid hypersecretion can be efficiently reduced by PPI therapy. PPI-therapy alone prevents symptoms related to Zollinger-Ellison syndrome.[4]

Insulin release may be reduced by diazoxide and verapamil.[39]

Isotope (^{90}Y, ^{177}Lu) labelled SS analogues are potent agents for the treatment of metastatic neuroendocrine tumors. These analogues are internalized by tumor cells (endoradiotherapy) resulting in high, local radiation.[59,60]

Pituitary Tumors

As fast growing and invasive pituitary tumors occur more frequently in MEN1 than with sporadic tumors, regular follow-up and aggressive treatment is necessary. Hormonally inactive, invasive macroadenomas should be treated with surgery and/or irradiation therapy.[18]

Prolactin-secreting tumors (prolactinomas) should be first treated with dopamine agonists (bromocriptin, quinagolide, cabergoline) that are effective not only in reducing prolactin secretion but may also lead to tumor shrinkage. In patients, who are nonresponsive to dopamin agonist therapy or are noncompliant, surgery or irradiation can be applied.[61]

Surgical intervention is the first choice for the treatment of GH-secreting tumors resulting in acromegaly, but SS analogues represent an effective treatment option for cases where surgery is contraindicated or unsuccessful. A GH receptor antagonist (pegvisomant) has been developed recently, but no studies have been performed in MEN1 patients to date.

ACTH-secreting tumors should also be treated by surgery. Drugs inhibiting steroid biosynthesis (mitotane, ketoconazole, etomidate etc.) can be used in cases, when surgery cannot be performed or was unsuccessful. As a last resort, bilateral adrenalectomy may be envisaged.

Surveillance

MEN1 patients should be regularly followed up after removal of MEN1-related tumors, as well. According to the guidelines of the National Comprehensive Cancer Network (NCCN), detailed history, physical examination, tumor markers, serum calcium, CT or MRI should be performed at 3 month postresection. Long term surveillance involves history taking, physical examination, measurement of tumor markers, serum calcium every 6 months in years 1-3 postresection, then annually. Somatostatin receptor scintigraphy and other imaging studies may also be performed.

Comments and Conclusion

MEN1 is a tumor syndrome with variable clinical appearance. With the exception of hyperparathyroidism that is an almost invariable manifestation of MEN1, other manifestations are quite variable. Most tumors related to MEN1 are benign, gastrinomas and thymic carcinoids, however, are mostly malignant and are the major mortality factors in MEN1. The life expectancy of MEN1 patients can be almost normal, provided that regular clinical, laboratory and imaging screening is performed and disease manifestations are treated properly.

Genetic diagnosis of MEN1 cannot be over-exaggerated. Patients with *MEN1* mutations need life-long surveillance, whereas no follow-up is necessary for nonaffected members of MEN1 families. In 10-15% of clinically unambiguous MEN1 cases, however, no mutation can be identified. It is possible that these cases are related to genetic alterations in unexamined regulatory regions or even in other genes. In MEN1 patients and families with undetectable *MEN1* gene mutations, clinical screening is of pivotal importance. MEN1 phenocopy also complicates diagnosis. MEN1 phenocopy can be a major problem in MEN1 families, as mild HPT and pituitary tumors may occur also in a sporadic setting in MEN1 family members. Genetic analysis for the known mutation can solve this problem. As the absence of *MEN1* gene mutation does not absolutely rule out MEN1, careful examination and follow-up is needed in sporadic MEN1 to differentiate it from MEN1 phenocopy.

References

1. Wermer P. Genetic aspects of adenomatosis of endocrine glands. Am J Med 1954; 16:363-371.
2. Brandi ML. Multiple endocrine neoplasia type 1. Rev Endocr Metab Disorders 2000; 1:275-282.
3. Thakker RV. Multiple endocrine neoplasia type 1. Endocrinol Metab Clin North Am 2000; 29:541-567.
4. Brandi ML, Gagel RF, Angeli A et al. Guidelines for diagnosis and therapy of MEN type 1 and type 2. J Clin Endocrinol Metab 2001; 86:5658-5671.
5. Karges W, Schaaf L, Dralle H et al. Clinical and molecular diagnosis of multiple endocrine neoplasia type 1. Langenbeck's Arch Surg 2002; 386:547-552.
6. Marx SJ, Stratakis CA. Multiple endocrine neoplasia. J Intern Med 2005; 257:2-5.
7. Lakhani VT, You YN, Wells SA. The multiple endocrine neoplasia syndromes. Annu Rev Med 2007; 58: 253-265.
8. Chandrasekharappa SC, Guru SC, Manickam P et al. Positional cloning of the gene for multiple endocirne neoplasia type 1. Science 1997; 276:404-407.
9. Marx SJ. Molecular genetics of multiple endocrine neoplasia types 1 and 2. Nature Rev Cancer 2005; 5:367-375.
10. Ellard S, Hattersley AT, Brewer CM et al. Detection of an MEN1 gene mutation depends on clinical features and supports current referral criteria for diagnostic molecular genetic testing. Clin Endocrinol 2005; 62:169-175.
11. Bilezikian JP, Brandi ML, Rubin M et al. Primary hyperparathyroidism: new concepts in clinical, densitometric and biochemical features. J Int Med 2005; 257:6-17.
12. Grama D, Skogseid B, Wilander E et al. Pancreatic tumors in multiple endocrine neoplasia type 1: clinical presentation and surgical treatment. World J Surg 1992; 16:611-618.
13. Doherty GM. Multiple endocrine neoplasia type 1: duodenopancreatic tumors. Surg Oncol, 2003; 12:135-143.
14. Gibril F, Venzon DJ, Ojeaburu JV et al. Prospective study of the natural history of gastrinoma in patients with MEN1: definition of an aggressive and a nonaggressive form. J Clin Endocrinol Metab 2001; 86:5282-5293.
15. Soga J, Yakuwa Y. Vipoma/diarrheogenic syndrome: a statistical evaluation of 241 reported cases. J Exp Clin Cancer Res 1998; 17:389-400.
16. Chastain MA. The glucagonoma syndrome: a review of its features and discussion of new perspectives. Am J Med Sci 2001; 321:306-320.
17. Soga J, Yakuwa Y. Somatostatinoma/inhibitory syndrome: a statistical evaluation of 173 reported cases as compared to other pancreatic endocrinomas. J Exp Clin Cancer Res 1999; 18:13-22.
18. Beckers A, Betea D, Socin HV et al. The treatment of sporadic versus MEN1-related pituitary adenomas. J Int Med 2003; 253: 599-605.
19. Waldmann J, Bartsch DK, Kann PH et al. Adrenal involvement in multiple endocrine neoplasia type 1: results of 7 years prospective screening. Langenbecks Arch Surg 2007; 392:437-443.
20. Teh BT, McArdle J, Chan SP et al. Clinicopathologic studies of thymic carcinoids in multiple endocrine neoplasia type 1. Medicine (Baltimore) 1997; 76:21-29.
21. Scarsbrook AF, Thakker RV, Wass JAH et al. Multiple endocrine neoplasia: spectrum of radiologic appearances and discussion of a multitechnique imaging approach. Radiographics 2006; 26:433-451.
22. Marini F, Falchetti A, del Monte F et al. Multiple endocrine neoplasia type 1. Orphanet J Rare Diseases 2006; 1:38-46.
23. Ashgarian B, Turner MI, Gibril F et al. Cutaneous tumors in patients with multiple endocrine neoplasm type 1 (MEN1) and gastrinomas: prospective study of frequency and development of criteria with high sensitivity and specificity for MEN1. J Clin Endocrinol Metab 2004; 89:5328-5336.
24. Asgharian B, Chen YJ, Patronas NJ et al. Meningiomas may be a component tumor of multiple endocrine neoplasia type 1. Clin Cancer Res 2004; 10:869-880.
25. Carty SE, Helm AK, Amico JA et al. The variable penetrance and spectrum of manifestations of multiple endocrine neoplasia type 1. Surgery 1998; 124:1106-1113.
26. Olufemi SE, Green JS, Manickam P et al. Common ancestral mutation in the MEN1 gene is likely responsible for the prolactinoma variant of MEN1 (MEN1 Burin) in four kindreds from Newfoundland. Hum Mutat 1998; 11:264-269.
27. Hao W, Skarulis MC, Simonds WF et al. Multiple endocrine neoplasia type 1 variant with frequent prolactinoma and rare gastrinoma. J Clin Endocrinol Metab 2004; 80:3776-3784.
28. Brandi ML, Falchetti A. Genetics of primary hyperparathyroidism. Urol Int 2004; 72 Suppl: 11-16.
29. Kassem M, Kruse TA, Wong FK et al. Familial isolated hyperparathyroidism as a variant of multiple endocrine neoplasia type 1 in a large Danish pedigree. J Clin Endocrinol Metab 2000; 85:165-167.
30. Miedlich S, Lohmann T, Schneyer U et al. Familial isolated primary hyperparathyroidism—a multiple endocrine neoplasia type 1 variant? Eur J Endocrinol 2001; 145:155-160.

31. Pannett AA, Kennedy AM, Turner JJ et al. Multiple endocrine neoplasia type 1 (MEN1) germline mutations in familial isolated primary hyperparathyroidism. Clin Endocrinol 2003; 58:639-646.
32. Simonds WF, James-Newton LA, Agarwal SK et al. Familial isolated hyperparathyroidism. Clinical and genetic characteristics of 36 kindreds. Medicine 2003; 81:1-26.
33. Dreijerink KMA, van Beek AP, Lentjes EGWM et al. Acromegaly in a multiple endocrine neoplasia type 1 (MEN1) family with low penetrance of the disease. Eur J Endocrinol 2005; 153:741-746.
34. Daly AF, Jaffrain-Rea ML, Ciccarelli A et al. Clinical characterization of familial isolated pituitary adenomas. J Clin Endocrinol Metab 2006; 91:3316-3323.
35. Burgess JR, Nord B, David R et al. Phenotype and phenocopy: the relationship between genotype and clinical phenotype in a single large family with multiple endocrine neoplasia type 1 (MEN1). Clin Endocrinol 2000; 53:205-211.
36. Hai N, Aoki N, Shimatsu A et al. Clinical features of multiple endocrine neoplasia type 1 (MEN1) phenocopy without germline MEN1 gene mutations: analysis of 20 Japanese sporadic cases with MEN1. Clin Endocrinol 2000; 52:509-518.
37. Smith JR, Oates ME. Radionuclide imaging of the parathyroid glands: patterns, pearls and pitfalls. Radiographics 2004; 24:1101-1115.
38. Jensen RT. Gastrinomas, advances in diagnosis and management. Neuroendocrinology 2004; (80 Suppl. 1):23-27.
39. Herder WW. Insulinoma. Neuroendocrinology 2004; (80 Suppl. 1):20-22.
40. Kao PC, Taylor RL, Service FJ. Proinsulin by immunochemiluminometric assay for the diagnosis of insulinoma. J Clin Endocrinol Metab 1994; 78:1048-1051.
41. Kaltsas GA, Besser GM, Grossman AB. The diagnosis and medical management of advanced neuroendocrine tumors. Endocr Rev 2004; 25:458-511.
42. Granberg D, Stridsberg M, Seensalu R et al. Plasma chromogranin A in patients with multiple endocrine neoplasia type 1. J Clin Endocrinol Metab 1999; 84:2712-2717.
43. Nobels FR, Kwekkeboom DJ, Coopmans W et al. Chromogranin A as a serum marker for neuroendocrine neoplasia: comparison with neuron-specific enolase and the alpha-subunit of glycoprotein hormones. J Clin Endocrinol Metab 1997; 82:2622-2628.
44. Perracchi M, Conte D, Gebbia C et al. Plasma chromogranin A in patients with sporadic gastro-entero-pancreatic neuroendocrine tumors or multiple endocrine neoplasia type 1. Eur J Endocrinol 2003; 148:39-43.
45. Giovanella L, La Rosa S, Ceriani L et al. Chromogranin A as serum marker of neuroendocrine tumors: comparison with neuron-specific enolase and correlation with immunohistochemical findings. Int J Biol Markers 1999; 14:160-166.
46. O'Shea D, Rohrer-Theurs AW, Lynn JA et al. Localization of insulinomas by selective intraarterial calcium injection. J Clin Endocrinol Metab 1996; 81:1623-1627.
47. Power N, Reznek RH. Imaging pancreatic islet cell tumours. Imaging 2002; 14:147-159.
48. Kaltsas G, Korbonits M, Heintz E et al. Comparison of somatostatin analog and metaiodobenzylguanidine radionuclides in the diagnosis and localization of advanced neuroendocrine tumors. J Clin Endocrinol Metab 2001; 86:895-902.
49. Kaltsas G, Rockall A, Papadogias D et al. Recent advances in radiological and radionuclide imaging and therapy of neuroendocrine tumors. Eur J Endocrinol 2004; 151:15-27.
50. Evanson EJ. Imaging the pituitary gland. Imaging 2002; 14:93-102.
51. Lenders JWM, Eisenhofer G, Mannelli M et al. Phaeochromocytoma. Lancet 2005; 66:665-675.
52. Geerdink EAM, van der Luijt RB, Lips CJM. Do patients with multiple endocrine neoplasia syndrome type 1 benefit from periodical screening? Eur J Endocrinol 2003; 149:577-582.
53. Balogh K, Patócs A, Majnik J et al. Genetic screening methods for the detection of mutations responsible for multiple endocrine neoplasia type 1. Mol Genet Metab 2004; 83:74-81.
54. Carling T, Udelsman R. Parathyroid surgery in familial hyperparathyroid disorders. J Intern Med 2004; 257:1-11.
55. Carling T. Multiple endocrine neoplasia syndrome: genetic basis for clinical management. Curr Opin Oncol 2004; 17:7-12.
56. Gibril F, Chen YK, Schrump DS et al. Prospective study of thymic carcinoids in patients with multiple endocrine neoplasia type 1. J Clin Endocrinol Metab 2003; 88:1066-1081.
57. Peacock M, Bilezikian JP, Klassen PS et al. The calcimimetic cinacalcet normalizes serum calcium in subjects with primary hyperparathyroidism. J Clin Endocrinol Metab 2005; 90:135-141.
58. Öberg K. Future aspects of somatostatin-receptor-mediated therapy. Neuroendocrinology, 2004; (80 Suppl. 1):57-61.
59. Waldherr C, Pless M, Maecke HR et al. The clinical value of (90Y-DOTA)-D-Phe1-Tyr3-octreotide (90Y-DOTATOC) in the treatment of neuroendocrine tumours: a clinical phase II study. Ann Oncol 2001; 12:941-945.

60. Kwekkeboom DJ, Teunissen JJ, Bakker WH et al. Radiolabeled somatostain analog (177-Lu-DOT-A⁰Tyr3)octreotate in patients with endocrine gastroenteropancreatic tumors. J Clin Oncol 2005; 23:2754-2762.
61. McCutcheon IE. Management of individual tumor syndromes: pituitary neoplasia. Endocrinol Metab Clin North Am 1994; 23:37-51.
62. Igaz P, Wiener Z, Szabó P et al. Functional genomics approaches for the study of sporadic adrenal tumour pathogenesis. Clinical implications. J Steroid Biochem Mol Biol 2006; 101:87-96.
63. Kulke MH, Mayer RJ. Carcinoid tumors. N Eng J Med 1999; 340:858-868.

CHAPTER 2

Genetic Background of MEN1:
From Genetic Homogeneity to Functional Diversity
Patrick Gaudray* and Günther Weber

Abstract

Multiple Endocrine Neoplasia Type 1 corresponds to a monogenic predisposition syndrome inherited as a dominant trait that affects a variety of endocrine tissues, in particular parathyroids, endocrine pancreas and anterior pituitary. It is caused by mutations in the MEN1 tumor suppressor gene that inactivate menin, the *MEN1* encoded protein. Menin is involved in cell cycle control and apoptosis through its participation in functional dynamics of chromatin and regulation of transcription. In addition, genetic investigations have implicated menin in the maintenance of genomic integrity. However, the role of menin does not—by far—end here. It plays (too) many roles in the control of cell life and normality, far beyond endocrine oncogenesis, making it unlikely that the function of menin can be deciphered only by genetic investigation. In this context, writing a chapter on the genetic background of MEN1 appears at the same time as a challenge and a paradox. A challenge as everything has been either already written on the topic or included in the present book. A paradox since genetics is simultaneously at the background and at the forefront of MEN1. Our attempts are thus more investigating new—as well as already open issues than delivering a catalog of *MEN1* gene mutations.

Introduction: The History of a Rare Endocrine Genetic Disease

Initially, multiple endocrine neoplasia Type 1(MEN1, OMIM 131100), then named multiple endocrine adenomatosis was described by Paul Wermer as a syndrome affecting the anterior pituitary, the parathyroids and the pancreatic islets in a family in which it was assumed to be "caused by a dominant autosomal gene with a high degree of penetrance".[1] Larsson et al have compared constitutional and tumor tissue genotypes of MEN1-related insulinomas and showed "that oncogenesis in these cases involves unmasking of a recessive mutation at (the MEN1) locus".[2] Hence, it became clear that the mutations that cause MEN1 are recessive although the syndrome is inherited as an autosomal dominant trait, similarly to most cancer predisposition genes. Despite an extensive variability in the expression of MEN1 between and within MEN1 families, genetic linkage could not evidence any genetic heterogeneity. The one *MEN1* gene was mapped in 1988 to the long arm of chromosome 11 (11q13) by family studies and found tightly linked the *PYGM* gene.[2] It turned out that it is only 70 kb telomeric to this gene that *MEN1* was identified by positional cloning ten years latter (Fig. 1).[3,4]

The *MEN1* gene, as it is known today, consists of 10 exons, spanning 9 kb of genomic sequence and encoding a nuclear protein of 610 aminoacids (menin). Menin does not reveal homologies to any other known protein. Protein sequence conservation indicates that *MEN1* is present throughout the animal world, from mollusks to *Homo sapiens* (Fig. 2). However, its functions might have

*Corresponding Author: Patrick Gaudray—CNRS UMR 6239 Faculté des Sciences et Techniques, Parc de Grandmont, avenue Monge, 37200 Tours, France.
Email: patrick.gaudray@univ-tours.fr

SuperMEN1: Pituitary, Parathyroid and Pancreas, edited by Katalin Balogh and Attila Patocs.
©2009 Landes Bioscience and Springer Science+Business Media.

evolved differently in human/mammals and other species. In fact, while it has been possible to generate homozygous mutations in Drosophila,[5] the double knock-out of *Men1* in the mouse is lethal early during embryogenesis.[6,7] In addition, no homozygous mutation has ever been found in humans either in families where consanguinity was suspected or even in a family where two affected children had received a defective allele from both parents.[8]

The menin protein and its multiple partners are detailed in other chapters of this book. From a genetic point of view, menin's complex partnership network might open the way to multiple tumor pathways in which those partners are being involved. Various mutations could affect the genes driving those pathways, although probably not as bona fide MEN1 alternative causes since no genetic heterogeneity has been observed in MEN1 families. Nevertheless, those genes might contribute to a fine tuning of MEN1 expressivity (gene modifiers).

On the Nature of the *MEN1* Gene

An interspecies genomic comparison in the immediate vicinity of the *MEN1* gene has not revealed conserved features apart from the sequences encoding menin. In vertebrates a MAP kinase gene, *MAP4K2*, is located immediately 3' of *MEN1* (Fig. 1). This genomic organization is not conserved in Drosophila. The intergenic region 5' of *MEN1* is seemingly not conserved. However, it is covered by repetitive elements both in man and in mouse over large distances (≥12 kb, Figs. 2, 3). These features developed independently since the repetitive elements, SINE/Alu in man and SINE/B2 in the mouse, are paralogs. SINE B2 elements can contain RNA polymerase II promoters[9] and repetitive elements may provide tissue-specific regulatory elements for promoter activity.[10] Alu sequences have recently been shown as the most reliable hallmarks for housekeeping genes.[11] Similarly, transcription factor binding sites have been found in Alu sequences that may be associated with early markers of development. In a genome-wide comparison, a correlation between the mouse B1 and human Alu densities within the corresponding upstream regions of orthologous genes has been observed,[12] as it is the case for the *MEN1* gene.

Northern blot analysis showed two RNA species (2.9 and 4.2 kb in size) in thymus and pancreas,[4] two target tissues for MEN1. We therefore suspected that *MEN1* might exist in more than one splice variant. By 3'-RACE only a single product was obtained, consistent with the theoretical polyadenylation site found in *MEN1*. 5' RACE gave a completely different picture. Different splice isoforms were detected, which hooked 6 alternative exons 1 (named a to f) to the first coding exon, exon 2 (Fig. 3). None of the possible exons 1 seems to alter the open reading frame.[13] However, the possibility exists that they could interfere differently with the translation of menin[14] and/or that they correspond to some sort of tissue specificity. Accordingly, nuclease protection assays revealed the splice forms were differently distributed in different cell lines (Lovisa Bylund, personal communication). For the shortest and apparently most abundant transcript containing e1b (as well as, probably e1a and e1c), the transcription initiation site was determined and was identical to the corresponding site in the mouse. For e1e and e1f, the 5' ends have not been determined, but would likely be located in the highly repetitive region. No probes could be developed to determine

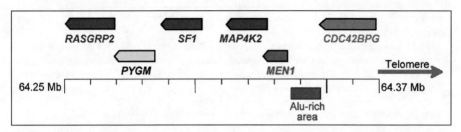

Figure 1. Genomic environment of the *MEN1* gene in the human genome. The information schematized here has been recovered from: http://www.ensembl.org/Homo_sapiens/contigview?region = 11&vc_start = 64250000&vc_end = 64370000.

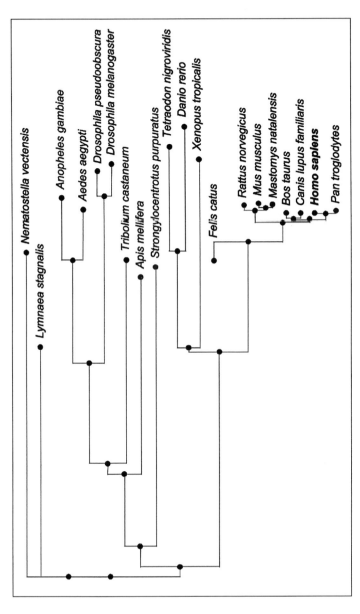

Figure 2. Conservation of menin sequence throughout evolution. The 610 aminoacid menin sequence has been blasted against the All nonredundant GenBank codons translations + RefSeq Proteins + Protein DataBase (PDB) + SwissProt databases and the results have been used to draw a phylogenic tree with the fast minimum evolution option of the NCBI program (http://www.ncbi.nlm.nih.gov/blast/Blast.cgi).

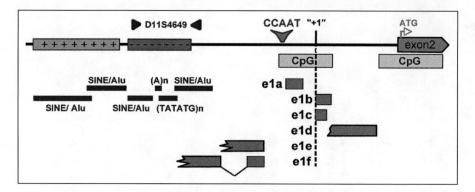

Figure 3. Organization of the 5' untranslated region of MEN1. The 1.5 kb genomic area located 5' to the translation start site of *MEN1* contains the only promoter identified to date and its regulatory regions. The – – and ++ boxes correspond to the upstream inhibitory and stimulatory regions, respectively. All exons which have been shown to be possibly spliced to exon 2, as well as repetitive elements and CpG islands are indicated.

whether the longest mRNA containing e1e and e1f, for instance, might correspond to the larger transcript seen in thymus and pancreas.

In the mouse two transcripts of 2.8 and 3.2 kb have been found in most tissues, though in variable amounts.[15] The larger transcript maintained the first intron and thus appeared similar to the e1d version in human. The four discrete transcripts identified in the mouse were expressed at varying time points and levels in the different tissues investigated.[16] Transcripts corresponding to human e1e or e1f were not detected in the mouse.

Altogether, these alternatively spliced transcripts should be studied with respect to their tissue specificity as they may represent a means of regulating *MEN1* expression at the translation level.[14] They would thus provide a fine-tuning of the menin protein level.

On the Regulation of the *MEN1* Gene

How the *MEN1* gene is regulated has been widely overlooked. However, it is quite an important matter, since it has to be determined whether cells bearing a heterozygous (hereditary) mutation in *MEN1* are still "normal". In this case, the so-called second event would happen as a stochastic event. An alternative possibility would be that the first mutation contributes in some ways to the "loss of heterozygosity". In fact, Payne and Kemp have raised the point that "... (the fact that) mutations in tumor suppressor genes can have haploinsufficient, as well as gain of function and dominant negative phenotypes has caused a reevaluation of the 'two-hit' model of tumor suppressor inactivation".[17] It is striking that most oncosuppressor genes whose mutations predispose to cancers are implicated in some ways in the maintenance of genome integrity.

MEN1 appears to be ubiquitously expressed and we have discussed above that the nature of its transcripts opens the issue of alternative—tissue specific? species that remain to be characterized. The composition of the upstream regulatory region and the constancy of its expression fits well with the behavior of a housekeeping gene involved in development. It should not be forgotten that there are still "classical" MEN1 families for which no mutations have been found within the coding region and for which an alteration within the regulatory regions should be considered.

From a quantitative point of view, the first level at which *MEN1* can be regulated is transcription. The region upstream the most abundant exon 1b is conserved (60%) between man and mouse and contains a putative CCAAT box, which, however, is not essential for the promoter activity.[18,19] The regions in man and mouse further upstream consist mostly of repetitive elements, without the features of a typical promoter but regulating the basic promoter activity when cloned into luciferase reporter vectors. We observed a direct feedback of menin that increased the promoter activity of this region after down regulation of menin by RNA interference. This may explain the normal levels of menin in cells from MEN1 patients and also that *MEN1* transcription level appears unchanged in tumor cells with loss of heterozygosity of *MEN1*.[20,21]

Two CpG islands are found around the transcriptional start site of exon 1b and further downstream from the first intron into the coding region of exon 2 (Fig. 3). However, their impact for transcriptional regulation is unknown. At present there is no evidence for epigenetic regulation of the *MEN1* gene. One study did not find any methylation in a part of the exon1-exon2 CpG island investigated.[22] Another found 36/108 pancreatic tumors methylated but without any significant correlation with the expression of menin.[23]

What Do We Learn from the Hereditary Mutations of the *MEN1* Gene?

Contrary to laboratory-engineered animal models, which, in case of knock-out mice represent single gross alterations of *Men1*, the diversity of human spontaneous mutations was hoped to shed light on the physiological mechanisms involved in endocrine carcinogenesis. Witnessing the efforts devoted to the search of *MEN1* mutations, more than 500 original reports were found on the basis of "MEN1 and mutations" keywords in the NCBI PubMed database. The most comprehensive MEN1 mutation databases can be found as part of the Universal Mutation Database (http://www.umd.be:2080/)[24] and the human gene mutation database at the Institute of Medical Genetics in Cardiff (http://www.hgmd.cf.ac.uk/). They include 353 and 392 independent *MEN1* mutations, respectively.

Most of the published *MEN1* mutations have been compiled in a recent survey that totalized 1133 independent germline mutations that consisted of 459 different mutations.[25] The same study surveyed also somatic mutations, i.e., mutation found in sporadic tumors and found 203 of them among which 167 different mutations. Altogether, it is total of 565 different mutations that have been found to affect possibly the *MEN1* gene. They are scattered over the all length of the known reading frame of the gene, although nine of them account for more than 23% of all mutations discovered to date. Taken together with the more recent results of Tham et al on a large series of MEN1 Swedish patients,[26] these studies widely agree on the following trends of *MEN1* gene mutations in the frame of the MEN1 syndrome:

- Less than 5% of the bona fide MEN1 do not present with identifiable mutations in *MEN1*.
- Probably ≈10% *MEN1* mutations arise de novo and may be transmitted to subsequent generations.[27]
- Approximately the two third of *MEN1* mutations lead to premature translation termination of the protein menin.
- No mutation or epimutation has ever been found in a potential regulatory region of the gene.
- Penetrance of MEN1 is almost complete after the seventh decade.[24,28]
- It has never been possible to draw a correlation between the nature of the mutation and the presentation of the disease, i.e., there is no correlation between genotype and phenotype. Even the long-lasting hypothesis that the so-called Burin-MEN1variant, clinical association of hyperparathyroidism, prolactinoma and carcinoids, was linked to specific mutations such as R460X has been ruled out in 2002.[24]

Importance of MEN1 in Endocrine Tumorigenesis

Functional loss of menin is not a prerequisite for the generation of endocrine tumors. Tumors with mutations of *MEN1* have their counterparts without mutations (or deletions at 11q13), which are undistinguishable phenotypically.

Extensive comparisons have been performed in sporadic parathyroid and pancreatic tumors, about 20-30% of which show mutations or loss of the *MEN1* gene.[29-31] In the case of pancreatic tumor, mutations or loss of expression of *MEN1* did not correlate with malignant development or the KI-67 proliferation index, in contrast to LOH on chr.1.[23,32] Deletions on 3p and 18q were found in familial and sporadic pancreatic endocrine tumors (PET) with a similar frequency, indicating that similar events lead to endocrine pancreatic carcinogenesis disregarding the presence or absence of functional menin.[33] Nonfunctional pancreatic tumors often showed deletion of the whole chromosome 11. However, this may be more relevant to the *ATM* gene present at 11q22 rather than *MEN1* at 11q13. In addition, no significance for prognosis could be observed.[34]

Despite these negative findings, there is little doubt that the loss of menin function contributes to the genesis of pancreatic tumors. In MEN1 patients, it has been shown that hyperplastic (precursor) lesions present in all patients consistently retained both 11q13 alleles.[35] The cause of hyperplasia development remains elusive, although it may be argued that the inherited *MEN1* mutation may have lead to a gene-dosage effect, at least in these patients. Nevertheless, whatever the mechanism of pancreatic hyperplasia, the loss of the second allele is a hallmark for neoplastic development and represent a "second step" in tumorigenesis.

Similarly, parathyroid tumors present with discrete patterns of deletions, including *MEN1* or not, all leading apparently to the same result.[29,30,36] We are not aware of any observation showing an increase of parathyroid hyperplasia in MEN1 patients similar to that described above for pancreas. In a microarray expression analysis, parathyroid carcinoma (as in the Hyperparathyroid-Jaw Tumor Syndrome caused by mutations of *HRPT2*), hyperplasia and adenoma formed different expression clusters.[37] This makes a progression model for parathyroid tumorigenesis rather unlikely.

In the case of typical MEN1, the occurrence of multiple tumors is a characteristic of the syndrome. For the parathyroids the rate of tumorigenesis might even be underestimated, since microdissection analysis has demonstrated polyclonality in parathyroid adenoma.[38] This is in contrast with a variant of MEN1, defined by sporadic tumors of both the parathyroids and pituitary. The prevalence of identified *MEN1* mutations in this variant is lower than in familial MEN1 (7% vs 90%).[39-43] In most of these families, the genetic cause of the disease is unknown, while in one case a mutation of $p27^{Kip1}/CDKN1B$, a gene known to be transcriptionally regulated by menin, has been observed.[44] The link between *MEN1* and $p27^{Kip1}/CDKN1B$ is still a matter of debate.[45]

Taken together, mutations studies suggest that functional loss of *MEN1* is less likely to affect the features and the malignancy of individual tumors, than to determine the tissue specificity and probability of neoplastic development in endocrine tissues. A means of driving the latter could be that menin is involved in the maintenance of genome stability.

Is MEN1 a Genome Instability Syndrome?

There has been too many discussion about how to categorize tumor suppressor genes, which usually ends up with the conclusion that a distinct classification is not possible. *RB1*, as a "classical" tumor suppressor gene, controls the cell cycle and its reintroduction may revert the tumorigenic phenotype of an *RB1*-negative cell.[46] Other tumor suppressor genes are directly involved in DNA repair, e.g., the *BRCA1/2* and Fanconi anemia genes as control elements of genome stability.[47] Their loss induces irreversible damage to the genome. However, this division into categories is too simple, since *RB1* loss can cause genome instability as well[48] and *BRCA1/2* genes have other functions than DNA repair, e.g., serving as transcription regulators.[49]

The function of menin as a classical tumor suppressor has been established both in tumor formation assays in the mouse[50] and in vitro.[51] However, menin has also been implicated in maintaining genome stability. In fact, significant chromosomal breakage has been observed specifically in lymphocytes from MEN1 patients, particularly after treatment with Diepoxybutane, a double-strand DNA damaging agent to which Fanconi anemia cells are also highly sensitive.[52-55] However, on the contrary to classical instability syndromes such as Fanconi anemia, MEN1 patients never present with homozygous *MEN1* mutations. Accordingly, homozygous mutations in the mouse model, have been proven lethal early during embryonic life. The phenotype observed in MEN1 patients respective to chromosome instability is subtle, in the context of a wild type allele. Hence, it is hardly convincing that MEN1 is a bona fide genome instability syndrome.[56]

In vitro cell cultures from MEN1 patients (i.e., genetically $MEN1^{+/-}$) show signs of chromosomal instability and premature centromere division when treated with mutagens.[52-57] An increase in the frequency of genetic loss has been observed in parathyroid tumors when MEN1 was mutated/deleted.[36] Similar conclusions have been drawn from LOH studies on pancreatic tumors.[33] It can be argued that a loss of genome stability should lead to higher malignancy than observed for the classical MEN1-associated tumors, with the exception of gastrinoma. More insight might be given by thymic foregut carcinoids that have been rarely been described but are the major cause of death in some MEN1 families. These highly malignant tumors occur predominantly in male patients, of which the majority were smokers or had previously been exposed to chemicals.[58,59] In line with the studies on cultured cells, this leads to the hypothesis that MEN1 might play its role in genome instability at the occurrence of particular genomic stress, such as the exposure to mutagens.

The animal model that has brought crucial information with respect to genome instability is *Drosophila melanogaster*. In these flies, the homozygous mutation of the MEN1 ortholog Mnn1 is not essential either in the regulation of cell growth or for differentiation. Mnn1 mutant flies show sensitivity to DNA damage and moderately to ionizing radiation, suggesting that menin is involved in the pathways leading to nucleotide excision repair.[5] In Drosophila as well, menin was required for an S-phase arrest and interacted with the transcription factor CHES1 upon exposure to ionizing irradiation, thus confirming its contribution to DNA damage response.[60]

Menin is localized to chromatin and nuclear matrix and its presence in nuclear matrix is dependent upon γ-irradiation. Menin has been shown to interact with FANCD2 and their association is also enhanced by γ-irradiation.[61] Among the Fanconi anemia-related proteins, FANCD2 has a central role between ATR that controls activation/monoubiquitinylation of FANCD2 and BRCA2 / RAD51 complexes acting more directly at the level of double-strand breaks repair. Interestingly, the binding of ATR to damaged DNA needs the interaction of ATRIP, which is dependent on RPA, another menin interacting protein.

Conclusion

When the MEN1 gene was still to be cloned, we expected the successful candidate gene to be "subtle". Most MEN1 tumors are benign and the cellular phenotype not excessively altered, as in a number of aggressive solid tumors in which crucial signal transduction pathways are primarily altered. The participation of menin in cell cycle control and apoptosis is in line with these expectations, although lots of questions remain as how menin precisely acts in these phenomena. In addition, genetic investigations have also implicated menin in the maintenance of genomic integrity. However, the role of menin does—by far—not end there.

Moreover, some lessons about the function of menin could not be taught solely by genetic investigations. What genetics did not tell us is that menin can play quite an opposite role as an oncogenic cofactor (and partner of MLL1) for MLL-associated myeloid transformation.[62] What it did not tell us either is the complexity of the oncogenic pathways and networks in which menin participates and which have been deciphered by functional studies. Most importantly, these have widened the playground for menin far beyond its link to endocrine tissues.

Based on those considerations, it seems most unlikely that "the" function of menin in tumorigenesis will ever be found. This protein plays just too many roles. A longtime focus on its role in endocrine glands has been a logical consequence of the MEN1 syndrome, though it is outdated. Menin is still a rather confusing, albeit central player in the control of the cell life and normality.

Acknowledgements

The authors want to acknowledge all the past and present members of their laboratories, as well as the French network on Multiple Endocrine Neoplasia Type 1 and Related Pathologies (CNRS GDR 2906) for many stimulating discussions and thoughts that found their way within the lines of this chapter. Our work is supported by the Ligue Nationale Française Contre le Cancer (Indre-et-Loire Committee).

References

1. Wermer P. Genetic aspects of adenomatosis of endocrine glands. Am J Med 1954; 16(3):363-371.
2. Larsson C, Skogseid B, Oberg K et al. Multiple endocrine neoplasia type 1 gene maps to chromosome 11 and is lost in insulinoma. Nature 1988; 332(6159):85-87.
3. Chandrasekharappa SC, Guru SC, Manickam P et al. Positional cloning of the gene for multiple endocrine neoplasia-type 1. Science 1997; 276(5311):404-407.
4. Lemmens I, Van de Ven WJ et al. Identification of the multiple endocrine neoplasia type 1 (MEN1) gene. The European Consortium on MEN1. Hum Mol Genet 1997; 6(7):1177-1183.
5. Busygina V, Suphapeetiporn K, Marek LR et al. Hypermutability in a Drosophila model for multiple endocrine neoplasia type 1. Hum Mol Genet 2004; 13(20):2399-2408.
6. Bertolino P, Radovanovic I, Casse H et al. Genetic ablation of the tumor suppressor menin causes lethality at mid-gestation with defects in multiple organs. Mech Dev 2003; 120(5):549-560.
7. Crabtree JS, Scacheri PC, Ward JM et al. A mouse model of multiple endocrine neoplasia, type 1, develops multiple endocrine tumors. Proc Natl Acad Sci USA 2001; 98(3):1118-1123.
8. Brandi ML, Weber G, Svensson A et al. Homozygotes for the autosomal dominant neoplasia syndrome (MEN1). Am J Hum Genet 1993; 53(6):1167-1172.
9. Ferrigno O, Virolle T, Djabari Z et al. Transposable B2 SINE elements can provide mobile RNA polymerase II promoters. Nat Genet 2001; 28(1):77-81.
10. Shephard EA, Chandan P, Stevanovic-Walker M et al. Alternative promoters and repetitive DNA elements define the species-dependent tissue-specific expression of the FMO1 genes of human and mouse. Biochem J 2007; 406(3):491-499.
11. Eller CD, Regelson M, Merriman B et al. Repetitive sequence environment distinguishes housekeeping genes. Gene 2007; 390(1-2):153-165.
12. Polak P, Domany E. Alu elements contain many binding sites for transcription factors and may play a role in regulation of developmental processes. BMC Genomics 2006; 7:133.
13. Khodaei-O'Brien S, Zablewska B, Fromaget M et al. Heterogeneity at the 5'-end of MEN1 transcripts. Biochem Biophys Res Commun 2000; 276(2):508-514.
14. van der Velden AW, Thomas AA. The role of the 5' untranslated region of an mRNA in translation regulation during development. Int J Biochem Cell Biol 1999; 31(1):87-106.
15. Stewart C, Parente F, Piehl F et al. Characterization of the mouse Men1 gene and its expression during development. Oncogene 1998; 17(19):2485-2493.
16. Forsberg L, Villablanca A, Valimaki S et al. Homozygous inactivation of the MEN1 gene as a specific somatic event in a case of secondary hyperparathyroidism. Eur J Endocrinol 2001; 145(4):415-420.
17. Payne SR, Kemp CJ. Tumor suppressor genetics. Carcinogenesis 2005; 26(12):2031-2045.
18. Fromaget M, Vercherat C, Zhang CX et al. Functional characterization of a promoter region in the human MEN1 tumor suppressor gene. J Mol Biol 2003; 333(1):87-102.
19. Zablewska B, Bylund L, Mandic SA et al. Transcription regulation of the multiple endocrine neoplasia type 1 gene in human and mouse. J Clin Endocrinol Metab 2003; 88(8):3845-3851.
20. Farnebo F, Teh BT, Kytola S et al. Alterations of the MEN1 gene in sporadic parathyroid tumors. J Clin Endocrinol Metab 1998; 83(8):2627-2630.
21. Wautot V, Khodaei S, Frappart L et al. Expression analysis of endogenous menin, the product of the multiple endocrine neoplasia type 1 gene, in cell lines and human tissues. Int J Cancer 2000; 85(6):877-881.
22. Chan AO, Kim SG, Bedeir A et al. CpG island methylation in carcinoid and pancreatic endocrine tumors. Oncogene 2003; 22(6):924-934.
23. Arnold CN, Sosnowski A, Schmitt-Graff A et al. Analysis of molecular pathways in sporadic neuroendocrine tumors of the gastro-entero-pancreatic system. Int J Cancer 2007; 120(10):2157-2164.
24. Wautot V, Vercherat C, Lespinasse J et al. Germline mutation profile of MEN1 in multiple endocrine neoplasia type 1: search for correlation between phenotype and the functional domains of the MEN1 protein. Hum Mutat 2002; 20(1):35-47.
25. Lemos MC, Thakker RV. Multiple endocrine neoplasia type 1 (MEN1): analysis of 1336 mutations reported in the first decade following identification of the gene. Hum Mutat 2008; 29(1):22-32.
26. Tham E, Grandell U, Lindgren E et al. Clinical testing for mutations in the MEN1 gene in Sweden: a report on 200 unrelated cases. J Clin Endocrinol Metab 2007; 92(9):3389-3395.
27. Bassett JH, Forbes SA, Pannett AA et al. Characterization of mutations in patients with multiple endocrine neoplasia type 1. Am J Hum Genet 1998; 62(2):232-244.
28. Pannett AA, Thakker RV. Multiple endocrine neoplasia type 1. Endocr Relat Cancer 1999; 6(4):449-473.
29. Carling T, Correa P, Hessman O et al. Parathyroid MEN1 gene mutations in relation to clinical characteristics of nonfamilial primary hyperparathyroidism. J Clin Endocrinol Metab 1998; 83(8):2960-2963.

30. Cetani F, Pardi E, Vignali E et al. MEN1 gene alterations do not correlate with the phenotype of sporadic primary hyperparathyroidism. J Endocrinol Invest 2002; 25(6):508-512.
31. Perren A, Komminoth P, Heitz PU. Molecular genetics of gastroenteropancreatic endocrine tumors. Ann N Y Acad Sci 2004; 1014:199-208.
32. Guo SS, Wu AY, Sawicki MP. Deletion of chromosome 1, but not mutation of MEN-1, predicts prognosis in sporadic pancreatic endocrine tumors. World J Surg 2002; 26(7):843-847.
33. Hessman O, Skogseid B, Westin G et al. Multiple allelic deletions and intratumoral genetic heterogeneity in men1 pancreatic tumors. J Clin Endocrinol Metab 2001; 86(3):1355-1361.
34. Rigaud G, Missiaglia E, Moore PS et al. High resolution allelotype of nonfunctional pancreatic endocrine tumors: identification of two molecular subgroups with clinical implications. Cancer Res 2001; 61(1):285-292.
35. Anlauf M, Perren A, Henopp T et al. Allelic deletion of the MEN1 gene in duodenal gastrin and somatostatin cell neoplasms and their precursor lesions. Gut 2007; 56(5):637-644.
36. Farnebo F, Kytola S, Teh BT et al. Alternative genetic pathways in parathyroid tumorigenesis. J Clin Endocrinol Metab 1999; 84(10):3775-3780.
37. Haven CJ, Howell VM, Eilers PH et al. Gene expression of parathyroid tumors: molecular subclassification and identification of the potential malignant phenotype. Cancer Res 2004; 64(20):7405-7411.
38. Lubensky IA, Debelenko LV, Zhuang Z et al. Allelic deletions on chromosome 11q13 in multiple tumors from individual MEN1 patients. Cancer Res 1996; 56(22):5272-5278.
39. Ellard S, Hattersley AT, Brewer CM et al. Detection of an MEN1 gene mutation depends on clinical features and supports current referral criteria for diagnostic molecular genetic testing. Clin Endocrinol (Oxf) 2005; 62(2):169-175.
40. Hai N, Aoki N, Matsuda A et al. Germline MEN1 mutations in sixteen Japanese families with multiple endocrine neoplasia type 1 (MEN1). Eur J Endocrinol 1999; 141(5):475-480.
41. Hai N, Aoki N, Shimatsu A et al. Clinical features of multiple endocrine neoplasia type 1 (MEN1) phenocopy without germline MEN1 gene mutations: analysis of 20 Japanese sporadic cases with MEN1. Clin Endocrinol (Oxf) 2000; 52(4):509-518.
42. Klein RD, Salih S, Bessoni J et al. Clinical testing for multiple endocrine neoplasia type 1 in a DNA diagnostic laboratory. Genet Med 2005; 7(2):131-138.
43. Ozawa A, Agarwal SK, Mateo CM et al. The parathyroid/pituitary variant of multiple endocrine neoplasia type 1 usually has causes other than p27^{Kip1} mutations. J Clin Endocrinol Metab 2007; 92(5):1948-1951.
44. Pellegata NS, Quintanilla-Martinez L, Siggelkow H et al. Germ-line mutations in p27^{Kip1} cause a multiple endocrine neoplasia syndrome in rats and humans. Proc Natl Acad Sci USA 2006; 103(42):15558-15563.
45. Polyak K. The p27^{Kip1} tumor suppressor gene: Still a suspect or proven guilty? Cancer Cell 2006; 10(5):352-354.
46. Huang HJ, Yee JK, Shew JY et al. Suppression of the neoplastic phenotype by replacement of the RB gene in human cancer cells. Science 1988; 242(4885):1563-1566.
47. Wang W. Emergence of a DNA-damage response network consisting of Fanconi anaemia and BRCA proteins. Nat Rev Genet 2007; 8(10):735-748.
48. Mayhew CN, Carter SL, Fox SR et al. RB loss abrogates cell cycle control and genome integrity to promote liver tumorigenesis. Gastroenterology 2007; 133(3):976-984.
49. Yoshida K, Miki Y. Role of BRCA1 and BRCA2 as regulators of DNA repair, transcription and cell cycle in response to DNA damage. Cancer Sci 2004; 95(11):866-871.
50. Ratineau C, Bernard C, Poncet G et al. Reduction of menin expression enhances cell proliferation and is tumorigenic in intestinal epithelial cells. J Biol Chem 2004; 279(23):24477-24484.
51. Hussein N, Casse H, Fontaniere S et al. Reconstituted expression of menin in Men1-deficient mouse Leydig tumour cells induces cell cycle arrest and apoptosis. Eur J Cancer 2007; 43(2):402-414.
52. Gustavson KH, Jansson R, Oberg K. Chromosomal breakage in multiple endocrine adenomatosis (types I and II). Clin Genet 1983; 23(2):143-149.
53. Itakura Y, Sakurai A, Katai M et al. Enhanced sensitivity to alkylating agent in lymphocytes from patients with multiple endocrine neoplasia type 1. Biomed Pharmacother 2000; 54 Suppl 1:187s-190s.
54. Scappaticci S, Maraschio P, del Ciotto N et al. Chromosome abnormalities in lymphocytes and fibroblasts of subjects with multiple endocrine neoplasia type 1. Cancer Genet Cytogenet 1991; 52(1):85-92.
55. Tomassetti P, Cometa G, Del Vecchio E et al. Chromosomal instability in multiple endocrine neoplasia type 1. Cytogenetic evaluation with DEB test. Cancer Genet Cytogenet 1995; 79(2):123-126.
56. Hecht F, Hecht BK. Unstable chromosomes in heritable tumor syndromes. Multiple endocrine neoplasia type 1 (MEN1). Cancer Genet Cytogenet 1991; 52(1):131-134.
57. Sakurai A, Katai M, Itakura Y et al. Premature centromere division in patients with multiple endocrine neoplasia type 1. Cancer Genet Cytogenet 1999; 109(2):138-140.

58. Teh BT. Thymic carcinoids in multiple endocrine neoplasia type 1. J Intern Med 1998; 243(6):501-504.
59. Teh BT, Zedenius J, Kytola S et al. Thymic carcinoids in multiple endocrine neoplasia type 1. Ann Surg 1998; 228(1):99-105.
60. Busygina V, Kottemann MC, Scott KL et al. Multiple endocrine neoplasia type 1 interacts with forkhead transcription factor CHES1 in DNA damage response. Cancer Res 2006; 66(17):8397-8403.
61. Jin S, Mao H, Schnepp RW et al menin associates with FANCD2, a protein involved in repair of DNA damage. Cancer Res 2003; 63(14):4204-4210.
62. Yokoyama A, Somervaille TC, Smith KS et al. The menin tumor suppressor protein is an essential oncogenic cofactor for MLL-associated leukemogenesis. Cell 2005; 123(2):207-218.

CHAPTER 3

Menin:
The Protein Behind the MEN1 Syndrome

Maria Papaconstantinou, Bart M. Maslikowski, Alicia N. Pepper and Pierre-André Bédard*

Abstract

The cloning of the *MEN1* gene in 1997 led to the characterization of menin, the protein behind the multiple endocrine neoplasia Type 1 syndrome. Menin, a novel nuclear protein with no homology to other gene products, is expressed ubiquitously. *MEN1* missense mutations are dispersed along the coding region of the gene but are more common in the most conserved regions. Likewise, domains of protein interaction often correspond to the more conserved segments of menin. These protein interactions are generally facilitated by multiple domains or encompass a large portion of menin. The exception to this rule is a small stretch of amino acids mediating the interaction of menin with the mSin3A corepressor and histone deacetylase complexes. The C-terminal region of menin harbors several nuclear localization signals that play redundant functions in the localization of menin to the nuclear compartment. The nuclear localization signals are also important for the interaction of menin with the nuclear matrix. Menin is the target of several kinases and a candidate substrate of the ATM/ATR kinases, implying a role for this tumor suppressor in the DNA damage response. Menin is highly conserved from *Drosophila* to human but is absent in the nematode and in yeast.

Introduction

Cloning of the gene for multiple endocrine neoplasia Type 1 (*MEN1*) in 1997 led to the identification of menin, the protein encoded by *MEN1*.[1,2] The 67 kDa menin is translated from a 2.8 kb transcript expressed ubiquitously and throughout mammalian development.[1,3-5] The *MEN1* gene is composed of 10 exons spanning a 9.0 kb region of genomic DNA (Fig. 1A).[1] Exon 1 is noncoding and accounts for most of the 5' UTR of the menin transcript. Exons 2 to 10 code for the 610 amino acid menin protein (id: AAC51229). A variant of 615 amino acids (id: AAC51230) was also identified in clones from a human leukocyte cDNA library and in a few EST clones.[1] The 615 amino acid variant, translated from an alternatively spliced transcript, contains an in-frame insertion of five amino acids at position 149 of the menin open reading frame. Several transcripts with different 5' UTR sequences have also been reported but the structure of the menin protein derived from these variants is not altered.[6] Little is known about the expression of the 615 amino acid menin and 5' UTR variants of the menin mRNA. One of the important conclusions of early molecular studies is that menin is a novel protein showing no homology to other gene products.

Menin Is a Nuclear Protein—Role of the C-Terminal Region

The nuclear localization of menin was confirmed by immunofluorescence, epitope tagging and western blotting analyses of subcellular fractions.[7] Deletion analysis identified the C-terminus

*Corresponding Author: Pierre-André Bédard—Department of Biology, McMaster University, Hamilton, Ontario, Canada. Email: abedard@mcmaster.ca

SuperMEN1: Pituitary, Parathyroid and Pancreas, edited by Katalin Balogh and Attila Patocs. ©2009 Landes Bioscience and Springer Science+Business Media.

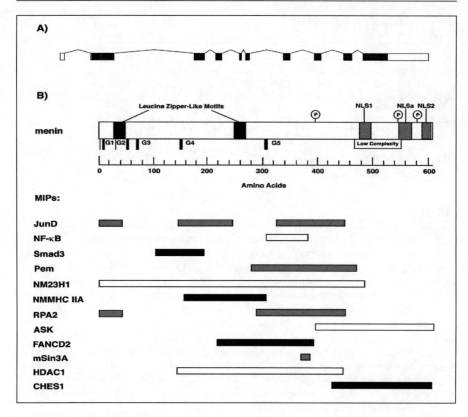

Figure 1. A) Exon-intron structure of the human *MEN1* gene. B) Schematic representation of functional domains of menin and binding sites of menin-interacting proteins (MIPs). Menin contains two leucine zipper-like motifs, two nuclear localization signals (NLS1 and NLS2) and an accessory NLS (NLSa). Five GTPase motifs (G1 to G5) are present in menin. A putative low complexity region spans amino acids 465 to 551. Menin is phosphorylated on two serine residues, Ser543 and Ser583, in 293T-cells. Phosphorylation of Ser394 in 293T-cells and M059K glioblastoma cells was observed in response to γ-irradiation and UV treatment, respectively. The regions of the menin sequence that have been implicated in the binding to different interacting proteins are indicated by black, white or gray bars under a schematic representation of the structure of menin.[12,20,25-27,45-50]

as the determinant of menin nuclear localization. Within this region, two independent nuclear localization signals (NLS1 and NLS2) are sufficient to target GFP or EGFP to the nuclear compartment.[7,8] Mutation analysis of NLS1 (amino acids 479-497) and NLS2 (amino acids 588-608) indicated that they play a redundant role in the nuclear localization of menin. More recently, La and coworkers reported that a third cluster of basic residues, located between amino acids 546 and 572, contributes to the nuclear localization of menin but is not sufficient to target GFP to the nucleus.[9] This region appears to function as an accessory signal (NLSa) in the nuclear localization of menin by NLS1 and NLS2 (Fig. 1B).

The localization of menin is consistent with the observation that several menin-interacting proteins are transcription factors, epigenetic regulators or proteins involved in DNA synthesis or repair (Chapter 5).[10] Menin associates with chromatin in vivo and binds dsDNA in vitro.[11,12] However, point mutations in any of the three NLSs abolish in vitro DNA binding without affecting the nuclear localization or association of menin with chromatin.[9] The same mutants are also deficient in repression of the *IGFBP2* promoter and inhibition of cell proliferation, implying an additional role for the NLSs apart from the nuclear localization of menin. Chromatin immunoprecipitation (ChIP) assays revealed that menin is recruited to the promoter region of the *IGFBP2* gene and that intact NLSs are required for this activity.[9] Menin also interacts with the 5' UTR region of the *caspase-8* locus in vivo and is capable of inducing the expression of this gene in mouse fibroblasts. Mutations in a single NLS also abolish the activation of the *caspase-8* gene by menin. Currently, there is no evidence that the DNA binding activity of menin is critical for recruitment to target loci. Menin binds DNA in a sequence-independent manner and, thus far, attempts at identifying specific DNA binding sites for menin have been unsuccessful.[13] Global genome analysis by ChIP assays indicates that menin is widely distributed in the genome, consistent with the notion that menin interacts with the phosphorylated carboxy terminal domain (CTD) of the RNA polymerase II large subunit, several transcription factors and epigenetic regulators.[14,15,16] It is possible that the recruitment of menin depends primarily on the interaction with a nuclear protein but that sequence-independent DNA binding by the positively charged NLSs plays a secondary role in this process.[9]

The importance of the C-terminal region of menin is highlighted by the fact that approximately 70% of the *MEN1* mutations are frameshift or nonsense mutations.[17,18] These mutations would disrupt the nuclear localization of menin and possibly other functions dependent on the presence of intact NLSs. Germline or somatic missense mutations have not been identified in NLS1 or NLS2 and only a few target the accessory region NLSa. This may be expected if the NLSs play a redundant role in the nuclear localization of menin but raises some questions regarding the significance of other activities dependent on the presence of intact NLSs.

The localization of menin to discrete nuclear foci has also been reported by separate groups in NIH 3T3 and HeLa cells.[12,19] Others, however, failed to observe this pattern of sub-nuclear localization using different cell types and antibodies, or a different approach such as GFP-tagging.[7] The association of menin with telomeres was described at meiotic prophase in mouse spermatocytes.[19] However, this pattern of menin localization was not observed in somatic cells and thus appears to be restricted to meiotic cells.

Leucine-Rich Domains in Menin

Two major approaches, yeast two-hybrid screening and proteomic analysis of co-immunoprecipitated proteins, have been employed to identify menin-interacting proteins and complexes (Chapter 5). The molecular analysis of these interactions indicated that large and/or multiple domains of menin are often required for association with its binding partners (Fig. 1B). The exception to this rule is a short stretch of centrally located amino acids (371-387) mediating the interaction of menin with the mSin3A corepressor and component of histone deacetylase complexes. The sixteen amino acid region, which includes tandemly repeated leucine residues, is predicted to form an amphipathic α-helix resembling the mSin3A-interacting domain (SID) of the Mad1 and Pf1 transcriptional repressors.[20-22] Point mutations within the SID region of menin abolish the interaction with mSin3A and the capacity of menin to repress the trans-activation function of JunD.[20]

Other leucine-rich regions have been the subject of investigation. Dreijerinck and collaborators showed recently that menin acts as a coactivator for nuclear receptor responsive genes such as *TFF1*.[23] In MCF7 breast carcinoma cells, menin was important for H3K4 trimethylation and ligand-dependent activation of this gene by the estrogen receptor α (ERα). Nuclear receptors interact with a leucine-rich motif in several coactivators (LXXLL). Since menin bound directly to the AF2 trans-activation domain of ERα in vitro these authors investigated the role of a conserved LLWLL motif (amino acids 263-267) in this process. Proteins encoded by clinically relevant *MEN1* mutations affecting the leucine-rich motif (L264P, L267P) were indeed deficient in the interaction with ERα and activation of the *TFF1* promoter. However, mutations in several residues outside the LLWLL motif also impaired this activity and thus menin may depend on a larger domain for the control of ERα-dependent gene expression.

The LLWLL motif is also part of a putative leucine zipper, one of two such candidate dimerization regions identified in menin (Fig. 1B).[24] However, the evidence of a role for these domains in protein dimerization, including menin homodimerization, is presently lacking.[4,24] Interestingly, menin does interact with the JunD bzip transcription factor but binds to the N-terminus and not the leucine zipper domain of JunD. Menin does not interact with other bzip members of the AP-1 family.[24]

GTPase Signature Motif

Menin associates with nm23H1/nucleoside diphosphate (NDP) kinase A, a multi-functional protein and candidate suppressor of tumor metastasis.[25] Results reported by Yamaguchi and co-investigators suggest that nm23H1 stimulates a latent GTPase activity in menin that is dependent on the interaction with nm23H1.[26] Menin was shown to bind GTP with low affinity but to hydrolyze GTP efficiently in association with nm23H1. A signature motif, conserved in GTPases and consisting of short stretches of a few amino acids grouped in five regions, was also identified in the N-terminal portion of menin. All regions (G1 to G5) are well conserved in mammals but not in more divergent species such as the zebrafish or *Drosophila melanogaster* (Fig. 2). Germline or somatic missense mutations in menin have been reported for the G4 region but are absent in other regions of the GTPase signature (Fig. 2). A subset of these *MEN1* missense mutants, analyzed by Yamaguchi and co-investigators, retained the GTPase activity.[26] Further studies are required to assess the significance of this GTPase activity and its relationship to the function of menin as a tumor suppressor.

Post-Translational Modification in Response to DNA Damage

Mouse embryo fibroblasts (MEFs) mutant for the *Men1* gene are hypersensitive to ionizing radiation and are deficient for a DNA damage-activated checkpoint.[27] Similar defects in cell cycle arrest were observed in *Drosophila* strains mutant for *Mnn1*, the *MEN1* homologue in the fruit fly, implying a conserved function for menin in the DNA damage response.[27] In agreement with this notion, two recent studies identified menin as a putative target of the ATM/ATR kinases.[28,29] A similar approach was employed in these studies wherein proteins phosphorylated in response to DNA damage were identified by immunoaffinity phosphopeptide isolation followed by sequence analysis using mass spectrometry. Using this approach, phosphorylation of Ser394 was detected in γ-irradiated 293T and UV-treated M059K glioblastoma cells.* Ser394 is located in a putative region of disordered secondary structure and is not conserved in other species (Figs. 2 and 3). Menin is also phosphorylated on Ser543 and Ser583 in 293T-cells but the kinase(s) responsible for phosphorylation of these residues and their function are presently unknown.[30] Ser583 is highly conserved and located in proximity of NLS2; Ser543 is present in mammalian menin but is not found in the zebrafish and *Drosophila* orthologues (Fig. 2). The association of menin with chromatin or the nuclear matrix is enhanced following UV treatment or γ-irradiation, suggesting that menin sub-nuclear localization or recruitment is regulated in response to DNA damage.[12,31] The treatment of UV-treated HEK293 cells with caffeine, an inhibitor of the ATR kinase, decreased menin localization to chromatin while overexpression of constitutively active CHK1 enhanced the association with chromatin. These results suggest that menin localization is regulated by an ATR-CHK1-dependent pathway in UV-treated cells.[31] The study of menin post-translational regulation is just beginning and additional sites of phosphorylation and modification are likely to be uncovered in the near future.

Conservation of Menin Structure, Protein Interactions and Function

MEN1 orthologues have been identified in vertebrates, sea urchin, snail and insects but are missing in the nematode and in yeast.[4] In vertebrates, the degree of protein identity to human menin ranges from 67% in the zebrafish to over 96% in mammalian species such as the mouse and rat (Fig. 2). Vertebrate species also share a similar exon/intron and overall structure of the *MEN1* gene, which consists of 10 exons with exon 1 being noncoding (Fig. 3A). Transcript variants with different 5′ UTRs, generated from alternative splicing of intron 1, have also been described in the mouse and rat.[5,32,33] The *Drosophila MEN1* gene (designated *Mnn1*) contains fewer exons but encodes a larger 763 amino acid protein 83 kDa in molecular weight.[34,35] Long stretches of amino acids located in the C-terminal portion of the protein and missing in vertebrate menin account for the larger size of *Drosophila* menin. Translation initiation is also predicted to occur at a different methionine codon, extending the N-terminus of *Drosophila* menin by 12 amino acids.[34] Cerrato and coworkers described the nuclear localization of menin in tissues of the third instar larva.[36] Putative nuclear localization signals are present at the C-terminus of *Drosophila* menin. A protein variant of 530 amino acids, lacking the C-terminal region of the 83 kDa protein, is predicted by the analysis of *Drosophila* cDNAs (id: NP 723252). We also described the expression of a 70 kDa heat shock-inducible form of menin in early embryos.[37] Little is known about the function of these menin variants in *Drosophila*.

Sequence divergence is found principally at the C-terminus of the protein (amino acids 465 to 551) in a proline-rich region of low sequence complexity that also accounts for much of the size difference in *Drosophila* menin (Fig. 2-3B). In mammals, the C-terminal region of menin is required for interaction with several proteins but fine mapping of this region is generally lacking (Fig. 1). Significantly, the majority (88%) of missense mutations identified to date are located in the N-terminal and "core" regions of menin (amino acids 1-447), i.e. in regions of greatest conservation (Figs. 2-3B). The overall identity of human and *Drosophila* menin is 46%. However, 65% of the 116 *MEN1* missense mutations identified thus far correspond to amino acids conserved in *Drosophila* menin.[18] This conservation reaches 82% in zebrafish, highlighting the importance of these amino acids in menin function. Prediction of secondary structure by the Russell/Linding definition (GlobProt)[38] identifies several putative regions of disordered structure, including the C-terminal region of low sequence complexity. *MEN1* missense mutations are lacking in regions predicted to be disordered (underlined in Fig. 2).

Unlike *Men1* in mammals, *Mnn1* function is not required for development in *Drosophila* and *Mnn1* mutant flies are fertile.[36,37,39-41] However, two different responses are affected by the absence of menin in the fruit fly. Defects in S-phase arrest have been described in response to ionizing radiation.[27] Mouse embryo fibroblasts mutant for *Men1* showed similar defects, implying a conserved function for menin in this response to DNA damage. Cell cycle arrest and viability were recovered in γ-irradiated flies overexpressing the forkhead family member Ches1. The human homologue, CHES1 (FOXN3), binds to human menin in vitro and co-immunoprecipitates in vivo, indicating a direct interaction between these proteins.[27] Whether or not this is also true for the *Drosophila* counterparts was not addressed in these studies.

We reported that *Mnn1* mutant strains are unable to mount a proper stress response. Developmental arrest and increased lethality were observed in response to heat shock, hypoxia, hyper-osmolarity and oxidative stress.[37] The expression of several heat shock proteins (HSPs) was impaired by the absence or over-expression of menin. *Mnn1* loss-of-function mutants expressed normal levels of HSP70 in the first 15 min of the heat shock response but were unable to sustain this expression beyond that point. In contrast, embryos over-expressing menin failed to down-regulate the expression of HSP70 upon return to the normal temperature. Thus, menin plays a role in the maintenance of HSP expression. In *Drosophila*, HSP70 expression depends on the H3K4 histone methyltransferase Trithorax (Trx) and components of the TAC1 chromatin modifying complex.[42] In mouse embryos and cells, menin interacts with histone methyltransferase (HMTase) complexes containing the Trithorax group proteins MLL2 or Ash2L. The action of these HMTases in the maintenance of *Hox* gene expression depends on their association with menin.[16,43,44] Recently, we observed that menin and Trithorax co-immunoprecipitate in protein lysates of *Drosophila* S2 cells, indicating that the interaction of menin with histone methyltransferase complexes has been conserved during evolution (our unpublished results[16,43]).

Menin interacts genetically with *Drosophila* Jun and Fos, components of the AP-1 transcription factor, but does not appear to bind directly to these proteins.[34,36] This contrasts with the direct interaction of menin and JunD described in human cells.[45] In general, the interaction of menin

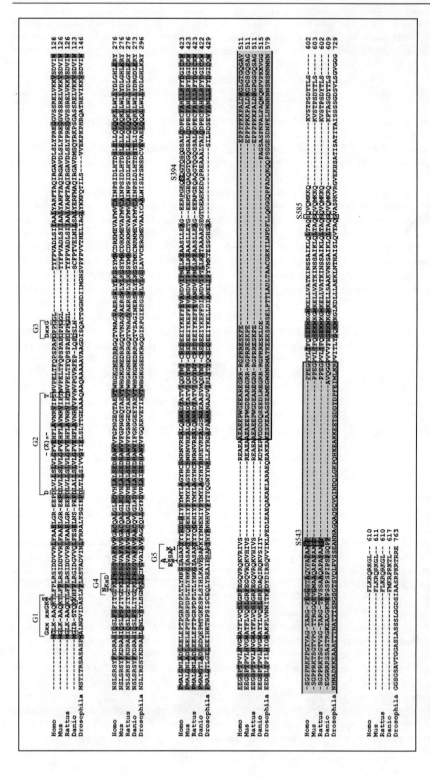

Figure 2. Figure legend viewed on following page.

Figure 2, viewed on previous page. Conservation of amino acids corresponding to missense point mutations in *MEN1*. Darkly shaded boxes indicate conserved residues in relation to known germline and somatic point mutations in human.[18] Underlined sequences show putative disordered secondary structure in the human sequence as predicted by GlobProt using the Russell/Linding definition.[38] Light grey box indicates a core low-complexity region as determined by multiple alignment of vertebrate sequences and predicted by SEG (our unpublished results).[51] Dashed-box regions show conserved putative GTPase motifs (G1-G5) determined by Yaguchi et al.[26] Consensus amino acid sequences are shown above. Solid boxes indicate serine phosphorylation sites Ser543 and Ser583 and the putative ATM and ATR phosphorylation site Ser394.[28-30,52] *Homo* indicates *Homo sapiens*; *Mus*, *Mus musculus*; *Rattus*, *Rattus norvegicus*; *Danio*, *Danio rerio*; and *Drosophila*, *Drosophila melanogaster*. A color version of this image is available at www.landesbioscience.com/curie

with known binding partners, identified in mammals, has not been investigated in *Drosophila* and other nonmammalian species.

Conclusion

Missense mutations have been identified along much of the menin open reading frame and do not cluster in "hot spots" pointing to domains of greater importance in the MEN1 syndrome. Regions of protein interaction are generally large or composed of multiple domains (Fig. 1B). Finer mapping of these regions will likely be important for the characterization of menin structure and function. Little is known about the dynamic of menin interaction with its binding partners.

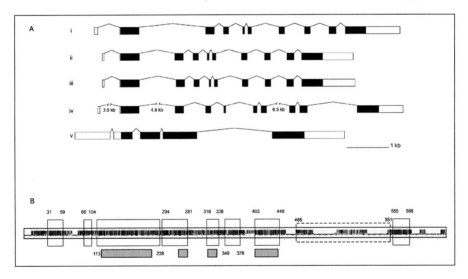

Figure 3. Panel A shows conservation of the exon-intron organization of the *MEN1* gene in (i) *Homo sapiens* (NC_000011.8) (ii) *Mus musculus* (NC_000085.5, AF109390) (iii) *Rattus norvegicus* (NC_005100.2, AB023400) (iv) *Danio rerio* (NC_007118.2) and (v) *Drosophila melanogaster* (NT_033779.4). Coding exons are indicated by black boxes and noncoding exons by white boxes. Introns are depicted as lines and approximate lengths are specified only for those introns too large to be drawn to scale. Panel B shows a schematic representation of the most highly conserved protein regions among five species (human, mouse, rat, zebrafish and fruit fly) menin homologues. Large boxes represent aligned regions with greatest degree of conservation in vertebrate species (calculated as the largest continuous aligned segments with lowest entropy). Gray boxes represent regions most conserved with *Drosophila* menin. Short functional motifs such as nuclear localization signals falling outside larger conserved regions are not indicated. The hatched box represents a putative core low complexity region in vertebrate sequences. Amino acid numbers refer to the human sequence.

The study of post-translational modifications of menin may shed some light on the mechanisms governing the interaction of menin with other proteins. A more extensive characterization of the menin-containing complexes will lead to the identification of additional binding partners and a better understanding of the role of menin in tumor suppression. The understanding of this role will continue to depend on multiple experimental approaches including the study of model systems, the characterization of domains of conserved protein interactions and the determination of the tri-dimensional structure of menin.

Acknowledgements

MP and ANP are recipients of a graduate scholarship from the Ontario Council on Graduate Studies and Natural Sciences and Engineering Research Council of Canada, respectively. Research in our laboratory is funded by grants form the Canadian Institutes of Health Research (MOP-10272 and MOP-84359).

References

1. Chandrasekharappa SC, Guru SC, Manickam P et al. Positional cloning of the gene for multiple endocrine neoplasia-type 1. Science 1997; 276(5311):404-407.
2. Lemmens I, Van de Ven WJ, Kas K et al. Identification of the multiple endocrine neoplasia type 1 (MEN1) gene. The european consortium on MEN1. Hum Mol Genet 1997; 6(7):1177-1183.
3. Wautot V, Khodaei S, Frappart L et al. Expression analysis of endogenous menin, the product of the multiple endocrine neoplasia type 1 gene, in cell lines and human tissues. Int J Cancer 2000; 85(6):877-881.
4. Chandrasekharappa SC, Teh BT. Functional studies of the MEN1 gene. J Intern Med 2003; 253(6):606-615.
5. Stewart C, Parente F, Piehl F et al. Characterization of the mouse Men1 gene and its expression during development. Oncogene 1998; 17(19):2485-2493.
6. Khodaei-O'Brien S, Zablewska B, Fromaget M et al. Heterogeneity at the 5'-end of MEN1 transcripts. Biochem Biophys Res Commun 2000; 276(2):508-514.
7. Guru SC, Goldsmith PK, Burns AL et al. Menin, the product of the MEN1 gene, is a nuclear protein. Proc Natl Acad Sci USA. 1998; 95(4):1630-1634.
8. La P, Schnepp RW, C DP et al. Tumor suppressor menin regulates expression of insulin-like growth factor binding protein 2. Endocrinology 2004; 145(7):3443-3450.
9. La P, Desmond A, Hou Z et al. Tumor suppressor menin: The essential role of nuclear localization signal domains in coordinating gene expression. Oncogene 2006; 25(25):3537-3546.
10. Agarwal SK, Kennedy PA, Scacheri PC et al. Menin molecular interactions: insights into normal functions and tumorigenesis. Horm Metab Res 2005; 37(6):369-374.
11. La P, Silva AC, Hou Z et al. Direct binding of DNA by tumor suppressor menin. J Biol Chem 2004; 279(47):49045-49054.
12. Jin S, Mao H, Schnepp RW et al. Menin associates with FANCD2, a protein involved in repair of DNA damage. Cancer Res 2003; 63(14):4204-4210.
13. La P, Yang Y, Karnik SK et al. Menin-mediated caspase 8 expression in suppressing multiple endocrine neoplasia type 1. J Biol Chem 2007; 282(43):31332-31340.
14. Agarwal SK, Impey S, McWeeney S et al. Distribution of menin-occupied regions in chromatin specifies a broad role of menin in transcriptional regulation. Neoplasia 2007; 9(2):101-107.
15. Scacheri PC, Davis S, Odom DT et al. Genome-wide analysis of menin binding provides insights into MEN1 tumorigenesis. PLoS Genet 2006; 2(4):e51.
16. Hughes CM, Rozenblatt-Rosen O, Milne TA et al. Menin associates with a trithorax family histone methyltransferase complex and with the hoxc8 locus. Mol Cell 2004; 13(4):587-597.
17. Marx SJ. Molecular genetics of multiple endocrine neoplasia types 1 and 2. Nat Rev Cancer 2005; 5(5):367-375.
18. Lemos MC, Thakker RV. Multiple endocrine neoplasia type 1 (MEN1): Analysis of 1336 mutations reported in the first decade following identification of the gene. Hum Mutat 2008; 29(1):22-32.
19. Suphapeetiporn K, Greally JM, Walpita D et al. MEN1 tumor-suppressor protein localizes to telomeres during meiosis. Genes Chromosomes Cancer 2002; 35(1):81-85.
20. Kim H, Lee JE, Cho EJ et al. Menin, a tumor suppressor, represses JunD-mediated transcriptional activity by association with an mSin3A-histone deacetylase complex. Cancer Res 2003; 63(19):6135-6139.
21. Yochum GS, Ayer DE. Pf1, a novel PHD zinc finger protein that links the TLE corepressor to the mSin3A-histone deacetylase complex. Mol Cell Biol 2001; 21(13):4110-4118.

22. Eilers AL, Billin AN, Liu J et al. A 13-amino acid amphipathic alpha-helix is required for the functional interaction between the transcriptional repressor Mad1 and mSin3A. J Biol Chem 1999; 274(46):32750-32756.
23. Dreijerink KM, Mulder KW, Winkler GS et al. Menin links estrogen receptor activation to histone H3K4 trimethylation. Cancer Res 2006; 66(9):4929-4935.
24. Agarwal SK, Guru SC, Heppner C et al. Menin interacts with the AP1 transcription factor JunD and represses JunD-activated transcription. Cell 1999; 96(1):143-152.
25. Ohkura N, Kishi M, Tsukada T et al. Menin, a gene product responsible for multiple endocrine neoplasia type 1, interacts with the putative tumor metastasis suppressor nm23. Biochem Biophys Res Commun 2001; 282(5):1206-1210.
26. Yaguchi H, Ohkura N, Tsukada T et al. Menin, the multiple endocrine neoplasia type 1 gene product, exhibits GTP-hydrolyzing activity in the presence of the tumor metastasis suppressor nm23. J Biol Chem 2002; 277(41):38197-38204.
27. Busygina V, Kottemann MC, Scott KL et al. Multiple endocrine neoplasia type 1 interacts with forkhead transcription factor CHES1 in DNA Damage Response. Cancer Res 2006; 66(17):8397-8403.
28. Matsuoka S, Ballif BA, Smogorzewska A et al. ATM and ATR substrate analysis reveals extensive protein networks responsive to DNA damage. Science 2007; 316(5828):1160-1166.
29. Stokes MP, Rush J, Macneill J et al. Profiling of UV-induced ATM/ATR signaling pathways. Proc Natl Acad Sci USA. 2007; 104(50):19855-19860.
30. MacConaill LE, Hughes CM, Rozenblatt-Rosen O et al. Phosphorylation of the menin tumor suppressor protein on serine 543 and serine 583. Mol Cancer Res 2006; 4(10):793-801.
31. Farley SM, Chen G, Guo S et al. Menin localizes to chromatin through an ATR-CHK1 mediated pathway after UV-induced DNA damage. J Surg Res 2006; 133(1):29-37.
32. Karges W, Maier S, Wissmann A et al. Primary structure, gene expression and chromosomal mapping of rodent homologs of the MEN1 tumor suppressor gene. Biochim Biophys Acta 1999; 1446(3):286-294.
33. Maruyama K, Tsukada T, Hosono T et al. Structure and distribution of rat menin mRNA. Mol Cell Endocrinol 1999; 156(1-2):25-33.
34. Guru SC, Prasad NB, Shin EJ et al. Characterization of a MEN1 ortholog from Drosophila melanogaster. Gene 2001; 263(1-2):31-38.
35. Maruyama K, Tsukada T, Honda M et al. Complementary DNA structure and genomic organization of drosophila menin. Mol Cell Endocrinol 2000; 168(1-2):135-140.
36. Cerrato A, Parisi M, Anna SS et al. Genetic interactions between Drosophila melanogaster menin and Jun/Fos. Dev Biol 2006; 298(1): 59-70.
37. Papaconstantinou M, Wu Y, Pretorius HN et al. Menin is a regulator of the stress response in drosophila melanogaster. Mol Cell Biol 2005; 25(22):9960-9972.
38. Linding R, Russell RB, Neduva V et al. GlobPlot: Exploring protein sequences for globularity and disorder. Nucleic Acids Res 2003; 31(13):3701-3708.
39. Crabtree JS, Scacheri PC, Ward JM et al. A mouse model of multiple endocrine neoplasia, type 1, develops multiple endocrine tumors. Proc Natl Acad Sci USA 2001; 98(3):1118-1123.
40. Bertolino P, Radovanovic I, Casse H et al. Genetic ablation of the tumor suppressor menin causes lethality at mid-gestation with defects in multiple organs. Mech Dev 2003; 120(5):549-560.
41. Busygina V, Suphapeetiporn K, Marek LR et al. Hypermutability in a Drosophila model for multiple endocrine neoplasia type 1. Hum Mol Genet 2004; 13(20):2399-2408.
42. Smith ST, Petruk S, Sedkov Y et al. Modulation of heat shock gene expression by the TAC1 chromatin-modifying complex. Nat Cell Biol 2004; 6(2):162-167.
43. Yokoyama A, Wang Z, Wysocka J et al. Leukemia proto-oncoprotein MLL forms a SET1-like histone methyltransferase complex with menin to regulate Hox gene expression. Mol Cell Biol 2004; 24(13):5639-5649.
44. Yokoyama A, Somervaille TC, Smith KS et al. The menin tumor suppressor protein is an essential oncogenic cofactor for MLL-associated leukemogenesis. Cell 2005; 123(2):207-218.
45. Agarwal SK, Guru SC, Heppner C et al. Menin interacts with the AP1 transcription factor JunD and represses JunD-activated transcription [In Process Citation]. Cell 1999; 96(1):143-152.
46. Heppner C, Bilimoria KY, Agarwal SK et al. The tumor suppressor protein menin interacts with NF-kappaB proteins and inhibits NF-kappaB-mediated transactivation. Oncogene 2001; 20(36):4917-4925.
47. Canaff LaH, G.N. Menin interacts directly with the TGF β signaling molecule Smad3 and MEN1 missense mutations within the interacting region have impaired TGF-β transcriptional activity. In: Stratatis CAaM SJ, ed. Ninth International workshop on multiple endocrine neoplasia. Bethesda: Blackwell Publishing, 2004; 255:721-722.
48. Lemmens IH, Forsberg L, Pannett AA et al. Menin interacts directly with the homeobox-containing protein Pem. Biochem Biophys Res Commun 2001; 286(2):426-431.

49. Sukhodolets KE, Hickman AB, Agarwal SK et al. The 32-kilodalton subunit of replication protein A interacts with menin, the product of the MEN1 tumor suppressor gene. Mol Cell Biol 2003; 23(2):493-509.
50. Schnepp RW, Hou Z, Wang H et al. Functional interaction between tumor suppressor menin and activator of S-phase kinase. Cancer Res 2004; 64(18):6791-6796.
51. Wootton JC. Non-globular domains in protein sequences: Automated segmentation using complexity measures. Comput Chem 1994; 18(3):269-285.
52. Obungu VH, Lee Burns A, Agarwal SK et al. Menin, a tumor suppressor, associates with nonmuscle myosin II-A heavy chain. Oncogene 2003; 22(41):6347-6358.

CHAPTER 4

Cellular Functions of Menin

Geoffrey N. Hendy*, Hiroshi Kaji and Lucie Canaff

Abstract

Since its discovery as a novel protein some 10 years ago, many cellular functions of menin have been identified. However, which ones of these relate specifically to menin's role as a tumor suppressor and which ones not remains unclear. Menin is predominantly nuclear and acts as a scaffold protein to regulate gene transcription by coordinating chromatin remodeling. It is implicated in both histone deacetylase and histone methyltransferase activity and, via the latter, regulates the expression of cell cycle kinase inhibitor and homeobox domain genes. TGF-β family members are key cytostatic molecules and menin is a facilitator of the transcriptional activity of their signaling molecules, the Smads, thereby ensuring appropriate control of cell proliferation and differentiation.

Introduction

The basic cell functions of menin will be reviewed. The focus will be on the role of menin in cell cycle regulation, DNA repair and chromatin remodeling. Whereas the primary structure of menin has been well conserved throughout evolution and orthologues are present in fruit fly, zebrafish and mouse, a menin homologue is apparently not present in nematodes and yeasts.

Cell Cycle

In the cell cycle, a gap (G1) phase is incorporated between nuclear division (M phase) and DNA synthesis (S phase); G2 phase occurs between S and M. Differentiated cells may exit G1 and enter a resting phase, G0. To enter S phase, activation of cyclin-dependent kinases (CDKs) is required. CDKs bind to a cyclin subunit to become catalytically competent and the cyclin-CDK complexes are tightly regulated. During G1 diverse signals are evaluated and on this basis the cell either enters S phase or enters G0 or undergoes apoptosis. The G2 phase is devoted to mending replication errors and ensuring that all is in order to proceed with mitosis. Oncogenic transformation is largely the result of malfunctions in these G1 and G2 mechanisms.

Before G1 and in the absence of mitogenic signals, CDK2 is kept inactive. In resting cells, E2F factors are bound to the retinoblastoma protein (Rb) or family members and inactivate them. Mitogens work by increasing D-type cyclins, which combine with CDK4 and CDK6 to phosphorylate and inactivate Rb. The E2Fs that are released activate transcription of genes encoding components supporting DNA replication.

Premature entry into S phase is prevented by inhibitors of the cyclin-CDK complexes. These cyclin dependent kinase inhibitors (CDKIs) include p15Ink4b, p16Ink4a, p18Ink4c, p21Cip1/WAF1, p27Kip1 and p57Kip2 and some may mediate cytostatic signals. On the other hand, mitogens can suppress the expression or location or activity of CDKIs. Mitogenic factors acting through receptor tyrosine kinases activate the Ras pathway to stimulate cell proliferation, growth

*Corresponding Author: Geoffrey N. Hendy—Calcium Research Laboratory, Rm. H4.67, Royal Victoria Hospital, 687 Pine Avenue West, Montreal, QC H3A 1A1, Canada.
Email: geoffrey.hendy@mcgill.ca

SuperMEN1: Pituitary, Parathyroid and Pancreas, edited by Katalin Balogh and Attila Patocs.
©2009 Landes Bioscience and Springer Science+Business Media.

and survival. In the GTP-bound state, the Ras-MEK-ERK cascade promotes CDK activation. ERK phosphorylates and stabilizes the transcription factor, c-Myc, that induces and inhibits expression of cyclin D1 and CDKIs, respectively.

The cytokine TGF-β and family members provide cytostatic signals that limit G1 progression and cell proliferation. TGF-β activates a membrane complex of serine/threonine kinase receptors that phosphorylates Smad2 and Smad3 that associate with Smad4 and the complex translocates to the nucleus where it regulates transcription in combination with coactivators and corepressors. A subset of the regulated genes is critical for arresting G1. In epithelial cells this involves induction of CDKIs and repression of c-Myc. Smad-2, -3 and -4 are considered as tumor suppressors and mutations in several components of the TGF-β signaling pathway are contributors to a wide variety of cancers.

Menin and the Cell Cycle

Menin is a nuclear protein in nondividing (interphase) cells[1,2] and it is only in mitosis when the nuclear membrane has dissolved that menin appears in the cytoplasm.[3] At this time it may be associated with cytoskeletal elements. Menin interacts with nonmuscle myosin II-A heavy chain (NMHC II-A) that mediates alterations occurring in cytokinesis and cell shape during cell division[4] and also interacts with the intermediate filament network proteins, glial fibrillary acidic protein (GFAP) and vimentin.[5] It is unclear whether this relocalization represents only a sequestering of menin before cytokinesis or that the protein is playing a functional role in this location during late mitosis.

Tumor suppressors BRCA1 and BRCA2 are poorly expressed in quiescent cells. By contrast, we found that menin protein is relatively well expressed in quiescent rat pituitary somatolactotrope GH4C1 cells at G0-G1.[2] The CDKIs, such as p21 and p27, are also well expressed in quiescent cells and we suggested that menin may function like these CDKIs (or regulate their expression). We found that the levels of menin transiently decrease as Rb protein becomes hyperphosphorylated as cells enter the cell cycle and then increase again as the cells enter S phase from the G1-S-phase boundary onward (see Fig. 1). Others have identified increases in menin mRNA at this time.[6] It is at this stage that expression of other tumor suppressors such as BRCA1, BRCA2 and p53 increases. Thus menin may play some role at the G1-S-phase checkpoint analogous to BRCA1, BRCA2 and p53. It is to be emphasized, however, that the relative changes in menin expression are modest (2-3-fold) relative to the marked changes in expression noted for other tumor suppressors over the course of the cell cycle. Changes in post-translational modification that might suggest alterations in activity throughout the cell cycle have yet to be examined.

Menin and the Retinoblastoma Protein

Studies with knockout mice have provided evidence that menin and Rb may operate in a common pathway to regulate cell proliferation.[7] Mice homozygous for either deletion of the Rb1 gene or the Men1 gene die in utero. Mice heterozygous for either deletion of the Rb1 gene[8] or the Men1 gene develop endocrine tumors.[9,10] In the Rb1+/- mice, intermediate pituitary and thyroid tumors occur frequently with less frequent development of pancreatic islet hyperplasia and parathyroid lesions. In the Men1+/- mice, pancreatic islet and anterior pituitary adenomas are common. In mice heterozygous for both Men1 and Rb1 deletion, pancreatic hyperplasia and tumors of the intermediate pituitary and thyroid occur at high frequency. The tumor spectrum in the double heterozygotes is a combination of those for the individual heterozygotes, with no decrease in age of onset. This suggests that menin and Rb function in a common pathway. This would be in contrast with studies of mice heterozygous for deletion of Rb1 and p53 that exhibit accelerated tumorigenesis and a broadening of the spectrum of tumor types observed to encompass all the types exhibited in the individual heterozygotes.[11]

Menin and CDK Inhibitors

p18-p27 double mutant mice, like Rb heterozygous mice, develop multiple endocrine neoplasias, including pituitary, thyroid, parathyroid and adrenal, providing evidence that p18 and p27

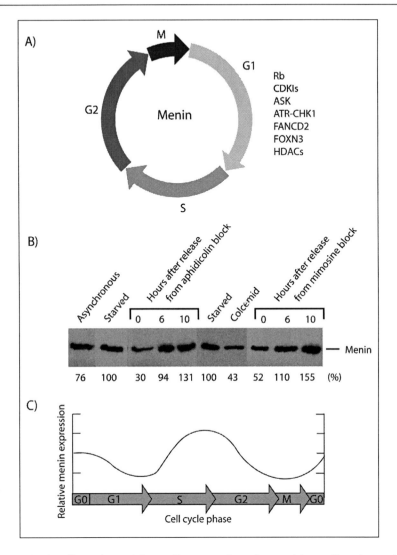

Figure 1. Menin affects the activity and/or expression of several key cell cycle regulators either by directly interacting with them or modulating transcription of their genes. A) G0/G1 to S phase transition; Menin blocks the transition from G0/G1 to S phase and appears to function in a common pathway with the retinoblastoma (Rb) protein. Menin is required for the expression of CDKIs such as p18 and p27 that maintain cells in a quiescent state. S phase: Menin inhibits cell proliferation by interacting with activator of S-phase kinase (ASK). Cell cycle checkpoints (G1 to S; and G2 to M) and DNA repair: Menin can bind DNA directly (like BRCA1) and functions in DNA repair via the ATR-CHK1, FANCD2 or FOXN3 (CHES1) pathways. For some of these, histone deacetylase (HDAC) complexes are also involved. For further details see text. B) Menin protein expression changes throughout the cell cycle. Rat somatolactotroph GH4C1 cells were serum starved for 24 h, cultured in complete media containing either aphidicolin or mimosine (G1-S block), or colcemid (G2-M block) for 24 h and then released from blockade by culture in complete media for the indicated times (h). For experimental details see reference 2. Relative expression levels (%) of menin protein were determined by SDS-PAGE and immunoblot of cell extracts. C) Relative expression of menin protein throughout the cell cycle with peak levels at the intra-S phase.

function by regulating Rb's tumor suppressor function. The types of endocrine tumors cover the spectrum seen in MEN1 and MEN2.[12,13] However, mice lacking p53 develop lymphomas and soft tissue sarcomas.[11] p53 functions as a checkpoint gene to monitor genomic integrity, whereas Rb functions to integrate mitogenic signals to determine whether the cell will enter S phase or not.

Menin directly regulates the expression of the CDKIs, p27 and p18. Menin does this by recruiting mixed lineage leukemia (MLL) protein complexes to their gene promoters and coding regions. Loss of function of either MLL or menin results in down-regulation of p27 and p18 expression and deregulated cell proliferation.[14] By use of a mouse model with heterozygous deletion of the *Men1* gene, it was shown that menin-dependent histone methylation maintained expression of CDKIs and prevented the formation of pancreatic islet tumors.[15] Therefore, menin is involved in cell proliferation control by an epigenetic mechanism. Excision of Men1 in mouse embryonic fibroblasts (MEFs) accelerated entry into S phase accompanied by increased CDK2 activity and decreased expression of p18 and p27. Complementation of menin-null cells with menin repressed S-phase entry.[16]

In stable Men1-deficient Leydig tumor cell lines reconstituted menin expression decreased cell proliferation with a block in transition from G0/G1 to S phase and an increase in apoptosis accompanied by increased p18 and p27 expression.[17] Tissue-specific inactivation of Men1 in neural crest precursor cells in the mouse leads to perinatal death with cleft palate and other cranial bone defects associated with decreased p27 expression.[18] The study demonstrated that menin functions in vivo during osteogenesis and is required for palatogenesis, skeletal rib formation and perinatal viability.

Therefore, a recurring theme arising from studies investigating the functions of menin either early in embryogenesis and fetal development, on the one hand and those examining roles that when lost in the growing or adult organism lead to endocrine (and other) neoplasia, on the other, has been the requirement of menin for the proper functioning of some of the CDKIs in cell cycle control.

Menin and GTPases

Ras-transformed murine fibroblasts (NIH3T3 cells) demonstrate increased proliferation, clonal formation in soft agar and tumor growth after inoculation into nude mice. Overexpression of menin in these cells caused them to partially revert to the phenotype of the parent NIH3T3 cells in vitro and in vivo.[19] The studies support menin acting as a tumor suppressor, although the activities being displayed by menin in these experiments may not necessarily involve a direct antagonism of the Ras pathway.

It has been suggested that nucleoside diphosphate (NDP) kinases might act as molecular switches to alter cell fate towards proliferation or differentiation in response to external signals.[20] The activity of the NDP kinases would be regulated by small molecular weight G proteins like Rad and Rac that function as guanosine triphosphate (GTP)ases. While menin is not of small molecular weight it has been shown to interact with the tumor suppressor NM23H1/NDP kinase. While neither protein has GTPase activity of its own, their interaction induces GTPase activity by menin.[21,22] It has been proposed that menin has several motifs similar to those in known GTPases although the homology is weak. It would be anticipated that the interaction with menin stimulates GEF activity of nm23 although this remains unknown. In addition NDP kinases function at the plasma membrane and evidence is lacking for menin being in the appropriate cellular localization to fulfil this function.

Menin and JunD

JunD is a member of the activator protein-1 (AP-1) transcription factor family and in contrast to other Jun and Fos proteins has antimitogenic activity. JunD negatively regulates fibroblast proliferation and antagonizes transformation by Ras in vitro and in vivo.[23] Of all the AP-1 family members, menin interacts only with JunD and represses its transcriptional activity by association with an mSin3A-histone deacetylase (HDAC) complex.[24-27] We had pointed out that it appeared paradoxical that one antimitogenic factor would reduce the activity of another.[2] We suggested that whereas menin is a regulator of JunD action, JunD may not be the main mediator of menin action. Further studies in fibroblasts have suggested that the nature of JunD can change depending upon whether it is bound by menin when it functions as a growth suppressor or it is not bound by menin in which case it acts as a growth promoter like other AP-1 family members.[28] However, the result of JunD-menin interaction may be cell-type specific as JunD has a differentiating effect in osteoblasts, an action that is inhibited by menin.[29]

Menin and Activator of S-Phase Kinase (ASK)

Targeted disruption of the *Men1* gene leads to enhanced cell proliferation, whereas complementation of menin-null cells with menin reduces cell proliferation. Menin interacts with activator of S-phase kinase (ASK), a component of the Cdc7/ASK kinase complex. The C-terminal domain of menin interacts with ASK. Wild-type menin completely represses ASK-induced cell proliferation although it does not affect the steady-state cell cycle profile of ASK-infected cells.[30] As menin itself represses basal cell proliferation, it is unclear whether menin's effects are occurring via its interaction with ASK or in an unrelated manner. Also, ASK itself did not alter the steady-state cell cycle profile. Disease-related C-terminal menin mutants that do not interact with ASK did not repress either ASK-induced or basal cell proliferation. From other studies it appears that ASK is in the nucleus.[31] The C-terminal menin deletion mutants would not gain access to the nucleus. Further studies need to be done to determine where in the cell any physical interaction between menin and ASK is taking place and what the exact functional link between the two proteins is.

Menin and TGF-β Family Members

In most mature tissues the cytokine TGF-β provides cytostatic signals that limit G1 progression and cell proliferation. Menin is a Smad3-interacting protein and is a facilitator of transcriptional activity of the Smads (see Fig. 2).[32] In anterior pituitary cells, inactivation of menin blocks TGF-β and activin signaling, antagonizing their proliferation-inhibitory properties.[33,34] In cultured parathyroid cells from uremic hemodialysis patients in which the menin signaling pathways are intact, menin inactivation achieved by menin antisense oligonucleotides leads to loss of TGF-β inhibition of parathyroid cell proliferation and parathyroid hormone (PTH) secretion (see Fig. 3). Moreover, TGF-β does not affect the proliferation and PTH production of parathyroid cells from MEN1 patients that were devoid of menin protein.[35,36] Antisense inhibition of menin in a rat duodenal crypt-like cell line increased cell proliferation with loss of cell-cycle arrest in G1 and increased expression of cyclin D1 and CDK4 and decreased expression of the TGF-β Type II receptor.[37] Hence, menin plays a critical role in mediating the cytostatic effects of TGF-β ligands.

During early embryogenesis and fetal development and in some adult mesenchymal cells, TGF-β and bone morphogenetic proteins (BMPs) play important roles. Homozygous Men1 inactivation in mice is embryonic lethal and the fetuses exhibit cranial and facial developmental defects. Cranial bones form by intramembranous ossification and menin may play an important role in this type of bone formation. In vitro, menin promotes the initial commitment of multipotential mesenchymal stem cells to the osteoblast lineage through interactions with the BMP-2 signaling molecules, Smad1/Smad5 and the key osteoblast regulator, Runx2, whereas the interaction of menin and the TGF-β signaling molecule, Smad3, inhibits later osteoblast differentiation by negatively regulating the BMP-Runx2 cascade.[38,39] The focus in these studies was predominantly on bone differentiation markers. However, it would also be important to extend these studies by evaluating menin's influence on cell cycle markers. In vivo, in the mouse, tissue-specific inactivation of Men1 in neural crest cells that contribute to cranial bones and the skeletal ribs and other tissues leads to defects in osteogenesis and perinatal death.[18] The fact that mice and humans heterozygous for loss of the *Men1* gene develop normally indicates haplosufficiency for all of menin's normal functions.[9,10] However, for some functions and in some cell types only, further reduction in menin results in developmental deficits.

The deregulation of the TGF-β family pathway has also been correlated with Men1 inactivation and altered cell growth in vivo by studying the Leydig cell tumors of heterozygous Men1 mutant mice.[40] In the cells of the tumors the anti-Mullerian hormone (AMH)/BMP pathway was impaired with reduced expression of AMH receptor Type 2, decreased expression of Smad1, -3, -4 and -5 and reduced BMP transcriptional activity. The expression of p18 and p27 was reduced and that of CDK4 increased. In other studies, it was noted that Men1-null MEFs demonstrate reduced expression of extracellular matrix proteins critical for organogenesis and that are induced by TGF-β.[41] TGF-β failed to stimulate expression of these proteins in the menin-null MEFs that also had poor responsiveness to TGF-β induced Smad3-mediated transcription.

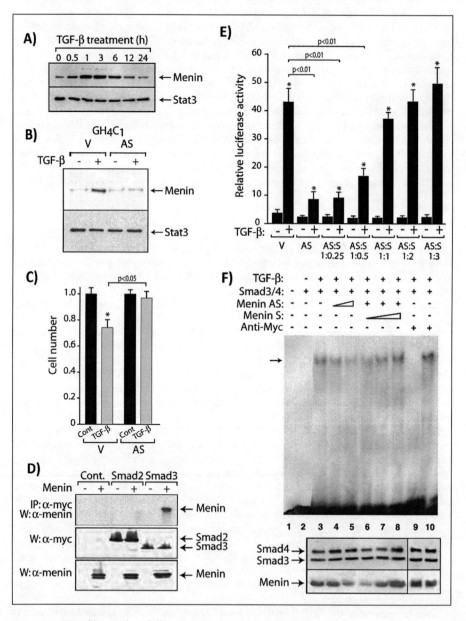

Figure 2. Legend viewed on following page.

Figure 2, viewed on previous page. Role of menin in TGF-β-mediated cell proliferation and gene transcription facilitated by Smad/DNA interaction. A) TGF-β stimulates menin expression. Serum-starved GH4C1 cells were cultured in TGF-β for the indicated times (h) and total cell menin levels were measured by immunoblot after SDS-PAGE with Stat3 as the protein loading control. B) Endogenous menin expression is suppressed by antisense menin cDNA. Serum-starved GH4C1 cells stably transfected with either vector alone (V) or antisense menin (AS) were cultured in the absence (-) or presence (+) of TGF-β for 1h. Menin expression was assessed by immunoblotting of cell extracts after SDS-PAGE. C) Antisense menin blocks the TGF-β-induced inhibition of pituitary cell proliferation. Serum-starved GH4C1 cells were cultured without (Cont.) or with TGF-β for 72 h and cell numbers were counted. D) Menin specifically binds the TGF-β signaling molecule, Smad3. Menin was transfected into COS7 cells with the indicated myc-tagged Smad2 or Smad3 constructs. Cell extracts were immunoprecipitated with anti-myc antibodies followed by SDS-PAGE and immunoblotting with anti-menin antibodies. Total cell expression of the Smads and menin was monitored. W, Western blot; IP, immunoprecipitation. E) Antisense menin inhibits TGF-β-mediated transcriptional responses. The TGF-β-responsive promoter-luciferase reporter construct, 3TP-Lux, was transfected into HepG2 cells together with empty vector (V) or antisense menin (AS) either alone or with sense menin (S) and the cells were stimulated (+) or not (-) with TGF-β. Relative luciferase activity was measured and the mean values are shown. F) Reduced menin expression disrupts Smad3 binding to DNA. GH4C1 cells were transfected with myc-Smad3 and flag-Smad4 in the absence (-) or presence (+) of antisense menin alone or with sense menin. Nuclear extracts were subjected to electromobility shift assay. The shifted band (arrow), the Smad/DNA complex, was decreased in intensity by antisense menin and restored by coexpression of sense menin and was completed abolished by anti-myc antibodies. Lane 10 represents an extract from cells transfected with untagged (rather than myc-tagged) Smad3. Menin and Smad3/4 in the nuclear extracts were monitored by immunoblot. For experimental details see reference 33.

Cell Cycle Checkpoints and DNA Repair

Tumor suppressors like BRCA1, BRCA2 and p53 play key roles in protection against genomic instability. They integrate with components of DNA damage checkpoints such as ataxia telangiectasia mutated kinase (ATM) and ATM and Rad3-related kinase (ATR) whose substrates mediate cell cycle arrest, DNA repair or cell death. Menin may also share some of these functions.

Menin and Genomic Instability

A role for menin in the maintenance of genomic stability is suggested. Chromosomal instability (increased chromosomal breakage) was observed in cultured lymphocytes and in fibroblasts derived from skin biopsies of MEN1 patients and hence heterozygous for an MEN1 mutation.[42] Peripheral lymphocytes from MEN1 patients displayed an increase relative to normal controls in premature centromere division after exposure to the alkylating agent diepoxybutane (DEB) that crosslinks DNA.[43-45] Menin-deficient MEFs were also moderately sensitive to DEB and displayed a high frequency of chromosomal aberrations after exposure to this agent.[46] Studies in Drosophila showed that flies mutant for the Men1 orthologue are hypersensitive to ionization radiation and are hypermutable.[46] A genome-wide loss of heterozygosity (LOH) screening of 23 pancreatic lesions from 13 MEN1 patients has shown multiple allelic deletions indicating that MEN1 pancreatic tumors fail to maintain DNA integrity and demonstrate signs of chromosomal instability.[47] However, in another study, no obvious chromosomal instability was observed in islet cells of Men1 knockout mice and tumors developed in the absence of chromosome or microsatellite instability.[48]

Menin, DNA Binding and ATR-CHK1 Pathway

Menin binds DNA and interacts with proteins implicated in DNA damage pathways. The canonical cellular response to UV-induced damage involves activation of the ATR kinase pathway. Following UV irradiation of human embryonic kidney (HEK293) cells, menin concentration in chromatin increased but was decreased by the ATR inhibitor, caffeine.[49] Transfection of constitutively active checkpoint kinase 1 (CHK1) increased chromatin-bound menin mimicking the effect of UV irradiation and implicating the involvement of an ATR-CHK1 dependent pathway.

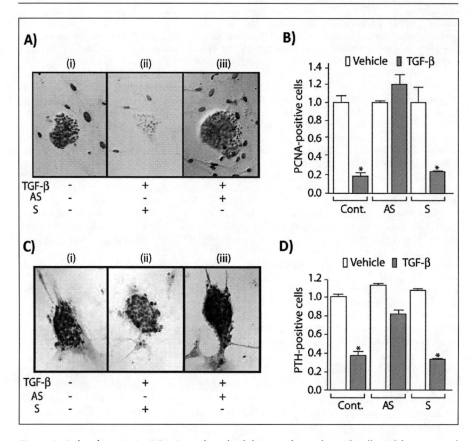

Figure 3. Role of menin in TGF-β-mediated inhibition of parathyroid cell proliferation and parathyroid hormone (PTH) production. Parathyroid cells (from patients with secondary hyperparathyroidism in which menin function is normal) were cultured in chamber slides without (-) or with (+) antisense (AS) or sense (S) menin oligos for 6 h. A) Cells were then cultured in fresh media without (-) or with (+) antisense or sense oligos without (-) or with (+) TGF-β for an additional 24 h and proliferating cell nuclear antigen (PCNA) immunocytochemistry performed. Representative views of cells cultured (i) without TGF-β or oligos, (ii) with TGF-β and sense oligos or (iii) with TGF-β and antisense oligos. B) Mean values for PCNA-positive cells for each group described in (A) without (vehicle) or with TGF-β. C) After the initial culture of cells for 6 h (described above) PTH immunocytochemistry was performed on some cells. Representative views of cells cultured as described under (B). D) Mean values of PTH-positive cells for each group described under (C) cultured without (vehicle) or with TGF-β. For experimental details see reference 35.

Similar to other tumor suppressors like BRCA1,[50] that are involved in DNA repair pathways, menin apparently directly binds dsDNA in a sequence independent way.[51] Amino acid sequences located towards the C-terminus of menin within the nuclear localization signals (NLS1 and NLS2) appear to be essential for this binding. The study was conducted in MEFs null for Men1 in which a failure to repress cell proliferation and cell cycle progression at the G2/M phase was noted. An anticipated effect at G1/S as occurs with BRCA1 depletion was not observed. However, the authors noted that the method of immortalization of the MEFs would have blocked the Rb pathway so that effects on this part of the cell cycle would not have been evaluated.

Menin and FANCD2

The FANCD2 protein is involved in DNA repair and mutations in it result in the inherited cancer syndrome, Fanconi's Anemia. Menin interacts with FANCD2; gamma-irradiation enhances the interaction and increases the accumulation of menin in the nuclear matrix.[52] These results suggest a role for menin, in cooperation with FANCD2, in DNA repair.

More recently, Marek et al[53] have compared the mutation frequency and spectra of Men1 and FANCD2 mutants in Drosophila. Men1 mutant flies were extremely prone to single base deletions within a homopolymeric tract, whereas FANCD2 mutants displayed large deletions. Neither overexpression nor loss of Men1 modified the interstrand crosslink (ICL) sensitivity of FANCD2 mutants. The different mutation spectra of Men1 and FANCD2 mutants together with lack of evidence for genetic interaction between these genes indicates that Men1 plays an essential role in ICL repair distinct from the Fanconi anemia genes.

Menin and RPA2

Menin directly binds to and colocalizes in the nucleus with the 32-kDa subunit (RPA2) of replication protein A (RPA), a heterotrimeric protein required for DNA replication, recombination and repair.[54] The interactive region was mapped to the N-terminal portion of menin and menin bound preferentially in vitro to free RPA2 rather than the RPA heterotrimer. Menin had no effect on RPA-DNA binding in vitro. However, menin antibodies coimmunoprecipitated RPA1 with RPA2 from HeLa extracts suggesting that menin binds to the RPA heterotrimer or a novel RPA1-RPA2-containing complex in vivo. The functional consequences of the interactions are unclear but it is possible that menin acts as a scaffold protein to bridge different components of a cross-link repair network. Menin could also have, by its association with HDACs, a more general effect on chromatin remodeling facilitating access of damaged DNA to the repair machinery.

Menin and FOXN3 (CHES1)

FOXN3 (CHES1), a member of the forkhead/winged-helix transcription factor family, was originally identified by its ability to suppress DNA damage sensitivity phenotypes in checkpoint-deficient yeast strains. By study of integrity of DNA damage checkpoints in mutant Drosophila lacking the Men1 orthologue and in MEFs deficient for Men1, biochemical and genetic interactions between menin and FOXN3 (CHES1) were demonstrated.[55] FOXN3 (CHES1) is part of a transcriptional repressor complex, that includes mSin3a, HDAC1 and HDAC2 and it interacts with menin in a DNA damage-responsive S-phase checkpoint pathway.

Chromatin Remodeling

Transcription factors bind to DNA in a sequence-specific manner and they recruit cofactors like chromatin-modifying complexes to regulate transcription of specific genes. In the nucleus, DNA is wrapped around histone proteins to form nucleosomes and the repeating nucleosomes form the chromatin fibres of chromosomes. Accessibility to chromosomal DNA is a prerequisite for gene transcription by RNA polymerases. Post-translational modifications of the tails of histones involving acetylation and methylation influence the status of nucleosomes and affect the recruitment of transcriptional cofactor complexes.

Menin and Transcriptional Regulation

Menin is implicated in the regulation of many genes via interaction with specific transcription factors and with the large subunit of RNA polymerase II.[56] Menin exerts a dual role, either as a repressor or as an activator depending upon the particular transcription factor involved. This might be explained by a model in which menin serves to link transcription regulation with chromatin modification.

Menin and Histone Deacetylase

The chromatin modification, histone acetylation, is correlated with activation of gene transcription. Histone deacetylation mediated by complexes of the general transcription repressor, Sin3A,

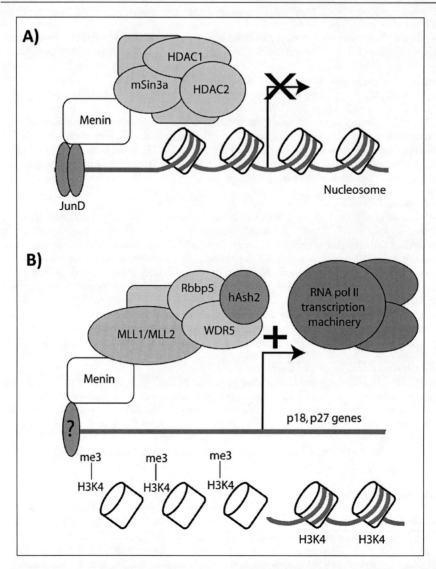

Figure 4. Menin regulates gene transcription. A) Menin interacts with JunD and represses JunD-mediated transcription by recruiting a histone-deacetylating complex comprising the mSin3A, HDAC1 and HDAC2 proteins. B) Menin is a component of MLL1 or MLL2 complexes with histone methyltransferase (H3K4) activity. Other proteins in the complex are WDR5, Rbbp5 and hASH2. Recruitment by an unknown DNA binding factor of menin and the histone methyl transferase complex leads to histone 3 lysine 4 trimethylation (H3K4-me3) coincident with the presence of RNA polymerase II and the basal transcription machinery at promoters such as those for the CDKI, p18 and p27, genes.

with HDACs is related to gene inactivation. Menin interacts with HDAC-mSin3A complexes to repress transcription of JunD (see Fig. 4A).[25,26]

Menin and Histone Methyltransferase

Menin, as a component of the MLL chromatin remodeling complexes, is involved in trimethylation of the fourth lysine (K) residue of histone H3 (H3K4 trimethylation) that is strongly associated with transcription activation.[56,57] By recruiting histone methyltransferase activity menin acts as a tumor suppressor by activating transcription of antiproliferative genes like those for CDKIs (see Fig. 4B).[14,15]

Menin binds to the N-terminal region of MLL1. Some but not all of the few MEN1 associated missense mutants tested fail to bind MLL.[56] So it remains unclear exactly how essential the loss of H3K4 methylation is in MEN1 tumorigenesis. Of interest, fusion proteins created by chromosomal translocations of MLL and other proteins in the complex have a causal role in several forms of leukemia. Menin is an essential cofactor for MLL-associated leukemogenesis.[58-60] This has prompted the suggestion that menin, usually considered as a tumor suppressor, under special circumstances may promote oncogenesis.

Homeobox-domain (HOX) genes that control the fundamental body plan during embryogenesis are targets for MLL. Consistent with menin being part of the MLL complexes, the protein is located at the promoters of some HOX genes and shown in some cases to regulate individual HOX gene expression in various cells including MEFs,[56] mouse bone marrow cells[59] and HeLa cells.[57,58,61] Deregulation of expression of many and a few HOX genes has been found in parathyroid tumors of patients with MEN1 and sporadic hyperparathyroidism, respectively.[62] The mechanisms underlying the altered HOX gene expression and whether it is causal of or coincidental to the tumorigenesis remains to be established.

Menin, Chromatin and Gene Expression

By combining chromatin immunoprecipitation (ChIP) assay with gene expression analysis, hundreds of menin-occupied chromatin regions were revealed.[63] The majority (67%) were at known genes (promoters, internal, 3' regions) with 33% located outside known gene regions. While this reinforces the notion of menin as a transcriptional regulator, menin likely acts in an indirect fashion (it may not bind DNA directly) by functioning as a scaffold protein within chromatin remodeling complexes. Knowledge of the participation of menin in regulating several genes has come from ChIP analysis and promoter-reporter transfection assays. Additional information has come from microarray analysis of a variety of Men1 expressing versus Men1 deleted cells. Interestingly, this has revealed that menin target genes in one tissue show minimal overlap with targets of other tissues.[63]

Genomic binding sites of menin and other proteins representing MLL complexes (see above) were mapped to several thousand gene promoters in three different cell types by ChIP coupled with microarray analyses.[61] While menin frequently colocalized with chromatin modifying complexes it also bound other promoters by other (unknown) mechanisms.

Conclusion

It is becoming clear that menin plays critical roles in embryogenesis and early fetal development for which functions menin appears to be haplosufficient. Menin is involved in organogenesis of neural tube, heart and craniofacial structures and hematopoiesis. In the adult, reductions in menin expression, under the influence of the hormone prolactin, has been implicated in the normal expansion of pancreatic islet β-cells that occurs in pregnancy to meet the increased insulin demand at this time.[64]

Acknowledgements

Work from our laboratories has been supported by the Canadian Institutes of Health Research (CIHR) (Grant MOP-9315 to G.N.H.) and the Kanzawa Medical Research Foundation (to H.K.) and the Ministry of Science, Education and Culture of Japan (Grant-in-aid 15590977 to H.K.).

References

1. Guru SC, Goldsmith PK, Burns AL et al. Menin, the product of the MEN1 gene, is a nuclear protein. Proc Natl Acad Sci USA 1998; 95:1630-4.
2. Kaji H, Canaff L, Goltzman D et al. Cell cycle regulation of menin expression. Cancer Res 1999; 59:5097-101.
3. Huang SC, Zhuang Z, Weil RJ et al. Nuclear/cytoplasmic localization of the multiple endocrine neoplasia type 1 gene product, menin. Lab Invest 1999; 79:301-10.
4. Obungu VH, Burns AL, Agarwal SK et al. Menin, a tumor suppressor, associates with nonmuscle myosin II-A heavy chain. Oncogene 2003; 22:6347-58.
5. Lopez-Egido J, Cunningham J, Berg M et al. Menin's interactions with glial fibrillary acidic protein and vimentin suggests a role for the intermediate filament network in regulating menin activity. Exp Cell Res 2002; 278:175-83.
6. Ikeo Y, Sakurai A, Suziki R et al. Proliferation-associated expression of the MEN1 gene as revealed by in situ hybridization: possible role of the menin as a negative regulator of cell proliferation under DNA damage. Lab Investig 2000; 80:797-804.
7. Loffler KA, Biondi CA, Gartside G et al. Lack of augmentation of tumor spectrum or severity in dual heterozygous Men1 and Rb1 knockout mice. Oncogene 2007; 26:4009-17.
8. Nitikin AY, Juarez-Perez MI, Li S et al. RB-mediated suppression of spontaneous multiple neuroendocrine neoplasia and lung metastases in Rb+/− mice. Proc Natl Acad Sci USA 1999; 96:3916-21.
9. Crabtree JS, Scacheri PC, Ward JM et al. A mouse model of multiple endocrine neoplasia type 1, develops multiple endocrine tumors. Proc Nat Acad Sci USA 2001; 98:1118-23.
10. Bertolino P, Tong WM, Galendo D et al. Heterozygous men1 mutant mice develop a range of endocrine tumors mimicking multiple endocrine neoplasia type 1. Mol Endocrinol 2003; 17:1880-92.
11. Harvey M, Vogel H, Lee EYH et al. Mice deficient in both p53 and Rb develop tumors primarily of endocrine origin. Cancer Res 1995; 55:1146-51.
12. Franklin DS, Godfrey VL, Lee H et al. CDK inhibitors p18INK4c and p27 Kip1 mediate two separate pathways to collaboratively suppress tumorigenesis. Genes Develop 1998; 12:2899-911.
13. Franklin DS, Godfrey VL, O'Brien DA et al. Functional collaboration between different cyclin-dependent kinase inhibitors suppress tumor growth with distinct tissue specificity. Mol Cell Biol 2000; 20:6147-58.
14. Milne TA, Hughes CM, Lloyd R et al. Menin and MLL cooperatively regulate expression of cyclin-dependent kinase inhibitors. Proc Natl Acad Sci USA 2005; 102:749-54.
15. Karnik SK, Hughes CM, Gu X et al. Menin regulates pancreatic islet growth by promoting histone methylation and expression of genes encoding p27Kip1 and p18INK4c. Proc Natl Acad Sci USA 2005; 102:14659-64.
16. Schnepp RW, Chen YX, Wang H et al. Mutation of tumor suppressor men1 acutely enhances proliferation of pancreatic islet cells. Cancer Res 2006; 66:5707-15.
17. Hussein N, Casse H, Fontaniere S et al. Reconstituted expression of menin in men1-deficient mouse leydig tumour cells induces cell cycle arrest and apoptosis. Eur J Cancer 2007; 43:402-14.
18. Engleka KA, Wu M, Zhang M et al. Menin is required in cranial neural crest for palatogenesis and perinatal viability. Dev Biol 2007; 311:524-37.
19. Kim YS, Burns AL, Goldsmith PK et al. Stable overexpression of MEN1 suppresses tumorigenicity of RAS. Oncogene 1999; 18:5936-42.
20. Kimura N, Shimada N, Ishijima Y et al. Nucleoside diphosphate kinases in mammalian signal transduction systems: recent development and perspective. J Bioenerg Biomembr 2003; 35:41-7.
21. Ohkura N, Kuhi M, Tsukada T et al. Menin, a gene product responsible for multiple endocrine neoplasia type 1, interacts with the putative tumor metastasis suppressor nm23. Biochem Biophys Res Commun 2001; 282:1206-10.
22. Yaguchi H, Ohkura N, Tsukada T et al. Menin, the multiple endocrine neoplasia type 1, gene product, exhibits GTP-hydrolysing activity in the presence of the tumor metastasis suppressor nm23. J Biol Chem 2002; 277:38197-204.
23. Pfarr CM, Mechta F, Spyrou G et al. Mouse JunD negatively regulates fibroblast growth and antagonizes transformation by ras. Cell 1994; 76:747-60.
24. Agarwal SK, Guru SC, Heppner C et al. Menin interacts with the AP1 transcription factor JunD and represses JunD-activated transcription. Cell 1999; 96:143-52.
25. Gobl AE, Berg M, Lopez-Egido LR et al. Menin represses JunD-activated transcription by a histone deacetylase-dependent mechanism. Biochim Biophys Acta 1999; 1447:51-6.
26. Kim H, Lee JE, Cho EJ et al. Menin, a tumor suppressor, represses JunD-mediated transcriptional activity by association with an mSin3A-histone deacetylase complex. Cancer Res 2003; 63:6135-9.
27. Yazgan O, Pfarr CM. Differential binding of the menin tumor suppressor protein to JunD isoforms. Cancer Res 2001; 61:916-20.

28. Agarwal SK, Novotny EA, Crabtree JS et al. Transcriptional factor JunD, deprived of menin, switches from growth suppressor to growth promoter. Proc Natl Acad Sci USA 2003; 100:10770-5.
29. Naito J, Kaji H, Sowa H et al. Menin suppresses osteoblast differentiation by antagonizing the AP-1 factor, JunD. J Biol Chem 2005; 280:4785-91.
30. Schnepp RW, Hou Z, Wang H et al. Functional interaction between tumor suppressor menin and activator of S-phase kinase. Cancer Res 2004; 64:6791-6.
31. Sato N, Sato M, Nakayama M et al. Cell cycle regulation of chromatin binding and nuclear localization of human Cdc7-ASK kinase complex. Gene Cell 2003; 8:451-63.
32. Hendy GN, Kaji H, Sowa H et al. Menin and TGF-β superfamily member signaling via the Smad pathway in pituitary, parathyroid and osteoblast. Horm Metab Res 2005; 37:375-9.
33. Kaji H, Canaff L, Lebrun JJ et al. Inactivation of menin, a Smad3-interacting protein, blocks transforming growth factor type-β signaling. Proc Natl Acad Sci USA 2001; 98:3837-42.
34. Lacerte A, Lee EH, Reynaud R et al. Activin inhibits pituitary prolactin expression and cell growth through Smads, pit-1 and menin. Mol Endocrinol 2004; 18:1558-1569.
35. Sowa H, Kaji H, Kitazawa R et al. Menin inactivation leads to loss of transforming growth factor-β inhibition of parathyroid cell proliferation and parathyroid hormone secretion. Cancer Res 2004; 64:2222-8.
36. Naito J, Kaji H, Sowa H et al. Expression and functional analysis of menin in a multiple endocrine neoplasia type 1 (MEN1) patient with somatic loss of heterozygosity in chromosome 11q13 and unidentified germline mutation of the MEN1 gene. Endocr 2006; 29:485-90.
37. Ratineau C, Bernard F, Poncet G et al. Reduction of menin expression enhances cell proliferation and is tumorigenic in intestinal epithelial cells. J Biol Chem 2004; 279:24477-84.
38. Sowa H, Kaji H, Canaff L et al. Inactivation of menin, the product of the multiple endocrine neoplasia type 1 gene, inhibits the commitment of multipotential mesenchymal stem cells into the osteoblast lineage. J Biol Chem 2003; 278:21058-69.
39. Sowa H, Kaji H, Hendy GN et al. Menin is required for bone morphogenetic protein 2- and transforming growth factor β-regulated osteoblastic differentiation through interaction with Smads and Runx2. J Biol Chem 2004; 279:40267-75.
40. Hussein N, Lu JL, Casse H et al. Deregulation of anti-Mullerian hormone/BMP and transforming growth factor-β pathways in leydig cell lesions developed in male heterozygous multiple endocrine neoplasia type 1 mutant mice. Endocrine-Related Cancer 2008; 15:217-27.
41. Ji Y, Prasad NB, Novotny EA et al. Mouse embryo fibroblasts lacking the tumor suppressor menin show altered expression of extracellular matrix protein genes. Mol Cancer Res 2007; 5:1041-51.
42. Scappaticci S, Maraschio P, del Ciotto N et al. Chromosome abnormalities in lymphocytes and fibroblasts of subjects with multiple endocrine neoplasia type 1. Cancer Genet Cytogenet 1991; 52:85-92.
43. Tomassetti P, Cometa G, Del Vecchio E et al. Chromosomal instability in multiple endocrine neoplasia type 1. Cytogenetic evaluation with DEB test. Cancer Genet Cytogenet 1995; 79:123-6.
44. Sakurai A, Katai M, Itakura Y et al. Premature centromere division in patients with multiple endocrine neoplasia type 1. Cancer Genet Cytogenet 1999; 109:138-40.
45. Itakura Y, Sakurai A, Katai M et al. Enhanced sensitivity to alkylating agent in lymphocytes from patients with multiple endocrine neoplasia type 1. Biomed Pharmacother 2000; 54(Suppl 1):187s-90s.
46. Busygina V, Suphapeetiporn K, Marek LR et al. Hypermutability in a drosophila model for multiple endocrine neoplasia type 1. Hum Mol Genet 2004; 13:2399-408.
47. Hessman O, Skogseid B, Westin G et al. Multiple allelic deletions and intratumoral genetic heterogeneity in MEN1 pancreatic tumors. J Clin Endocrinol Metab 2001; 86:1355-61.
48. Scacheri PC, Kennedy AL, Chin K et al. Pancreatic insulinomas in multiple endocrine neoplasia, type I knockout mice can develop in the absence of chromosome instability or microsatellite instability. Cancer Res 2004; 64:7039-44.
49. Farley SM, Chen G, Guo S et al. Menin localizes to chromatin through an ATR-CHK1 mediated pathway after UV-induced DNA damage. J Surg Res 2006; 133:29-37.
50. Paull TT, Cortez D, Bowers B et al. Direct DNA binding by brca1. Proc Natl Acad Sci USA 2001; 98:6086-91.
51. La P, Silva AC, Hou Z et al. Direct binding of DNA by tumor suppressor menin. J Biol Chem 2004; 279:49045-54.
52. Jin S, Mao H, Schnepp RW et al. Menin associates with FANCD2, a protein involved in repair of DNA damage. Cancer Res 2003; 63:4204-10.
53. Marek LR, Kottemann MC, Glazer PM et al. MEN1 and FANCD2 mediate distinct mechanisms of DNA crosslink repair. DNA Repair (Amst) 2008; 7:476-86.
54. Sukhodolets KE, Hickman AB, Agarwal SK et al. The 32-kilodalton subunit of replication protein A interacts with menin, the product of the MEN1 tumor suppressor gene. Mol Cell Biol 2003; 23:493-509.

55. Busygina V, Kottemann MC, Scott KL et al. Multiple endocrine neoplasia type 1 interacts with forkhead transcription factor CHES1 in DNA damage response. Cancer Res 2006; 66:8397-403.
56. Hughes CM, Rozenblatt-Rosen O, Milne TA et al. Menin associates with a trithorax family histone methyltransferase complex and with the hoxc8 locus. Mol Cell 2004; 13:587-97.
57. Yokoyama A, Wang Z, Wysocka J et al. Leukemia proto-oncoprotein MLL forms a SET1-like histone methyltransferase complex with menin to regulate hox gene expression. Mol Cell Biol 2004; 24:5639-49.
58. Yokoyama A, Somerville TC, Smith KS et al. The menin tumor suppressor protein is an essential oncogenic cofactor for MLL-associated leukemogenesis. Cell 2005; 123:207-18.
59. Chen YX, Yan J, Keeshan K et al. The tumor suppressor regulates hematopoiesis and myeloid transformation by influencing hox gene expression. Proc Natl Acad Sci USA 2006; 103:1018-23.
60. Caslini C, Yang Z, El-Osta M et al. Interaction of MLL amino terminal sequences with menin is required for transformation. Cancer Res 2007; 67:7275-83.
61. Scacheri PC, Davis S, Odom DT et al. Genome-wide analysis of menin binding provides insights into MEN1 tumorigenesis. PloS 2006; 2:e51.
62. Shen H-CJ, Rosen JE, Yang LM et al. Parathyroid tumor development involves deregulation of homeobox genes. Endocrine-Related Cancer 2008; 15:267-75.
63. Agarwal SK, Impey S, McWeeney S et al. Distribution of menin-occupied regions in chromatin specifies a broad role of menin in transcriptional regulation. Neoplasia 2007; 9:101-7.
64. Karnik SK, Chen H, McLean GW et al. Menin controls growth of pancreatic β-cells in pregnant mice and promotes gestational diabetes mellitus. Science 2007; 318:806-9.

CHAPTER 5

The Role of Menin in Hematopoiesis

Ivan Maillard and Jay L. Hess*

Abstract

In the hematopoietic system, menin was found to interact with MLL, a large protein encoded by the mixed linage leukemia gene that acts as a histone H3 methyltransferase. The *MLL* gene is a recurrent target for translocations in both acute myeloid and acute lymphoid leukemias. MLL gene rearrangements involve a variety of translocation partners, giving rise to MLL fusion proteins whose transforming ability is mediated through upregulated expression of *Homeobox* (*Hox*) genes as well as other targets. Recent work indicates that menin is an essential partner of MLL fusion proteins in leukemic cells and that it regulates normal hematopoiesis. In the absence of menin, steady-state hematopoiesis is largely preserved; however, menin-deficient hematopoietic stem cells are markedly deficient in situations of hematopoietic stress, such as during recovery after bone marrow transplantation. In leukemias driven by MLL fusion proteins, menin is essential for transformation and growth of the malignant cells. Thus, menin-MLL interactions represent a promising therapeutic target in leukemias with MLL rearrangements.

Introduction

Menin Is Associated with MLL, a Histone Methyltransferase Rearranged in Leukemia

Recognition of the role of menin in normal and neoplastic hematopoiesis arose from studies of the mixed lineage leukemia gene *MLL*, the mammalian homolog of *Drosophila* trithorax. MLL rearrangements are a common cause of both acute lymphoid and myeloid leukemias (Fig.1). Among lymphoid leukemias the most common MLL rearrangements are the t(4; 11) and t(11; 19) translocations, which are associated with pro-B-cell leukemias that express *MLL-AF4* and *MLL-ENL* respectively. The most common MLL rearrangements in acute myeloid leukemias include the t(9; 11), t(11; 19) and t(10; 11), which express *MLL-AF9, MLL-ELL* and *MLL-AF10* respectively. MLL rearrangements are also common in secondary acute lymphoid and myeloid leukemias arising following therapy with topoisomerase inhibitors such as etoposide.[1] In all, more than 50 different translocations have been identified.

MLL (3968 aa) is proteolytically cleaved into two fragments before entering the nucleus.[2,3] The amino terminus of (MLL^N) is a 300 kD protein that directly targets MLL to specific chromosomal sites including promoters and coding regions of *Hox* genes. These sequences span a short evolutionarily conserved N-terminal domain (NTD), three AT hooks, which bind the minor groove of DNA and a nonenzymatic DNA methyltransferase homology (DNMT) region that is pivotal for MLL binding to unmethylated CpG-rich DNA.[4,5] The C terminus of MLL^N contains several regions with high homology to trx. These include three cysteine-rich zinc finger domains (termed

---

*Corresponding Author: Jay L. Hess—Department of Pathology, University of Michigan Medical School, 5249 Medical Sciences 1, 1301 Catherine Avenue, Ann Arbor, Michigan 48105 USA. Email: jayhess@umich.edu

SuperMEN1: Pituitary, Parathyroid and Pancreas, edited by Katalin Balogh and Attila Patocs. ©2009 Landes Bioscience and Springer Science+Business Media.

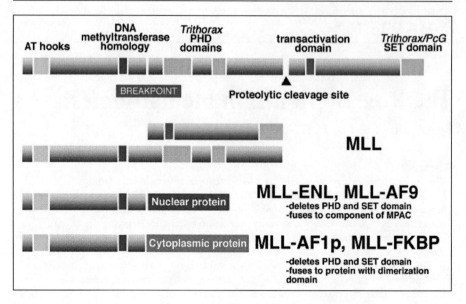

Figure 1. Schematic of MLL and the two general types of MLL fusion proteins. The most common translocations fuse MLL to nuclear translocation partners with transcriptional activating activity. Many of these are components of the MLL Partner Activating Complex (MPAC) (see text and Fig. 2). Less commonly, MLL is fused to a translocation partner that transforms through dimerization of the truncated MLL molecule. MLL-FKBP is an experimental fusion protein that transforms only in the presence of synthetic dimerizer. MLL and both classes of MLL fusion proteins interact with menin via sequences in the extreme amino terminus.

PHD for plant homeodomain) which flank an imperfect bromodomain, a domain implicated in binding to acetylated histones.[2,3,6-8]

The 180 kD MLL^C peptide noncovalently associates with MLL^N and has potent transcriptional activating activity.[2,3] Transcriptional activation by MLL involves a concerted series of histone modifications mediated by MLL^C. The MLL SET domain has intrinsic histone methyltransferase activity specific for histone H3 lysine 4 and that this plays a major role in transcriptional activation.[9,10] In addition, histone acetylation also contributes to transcriptional activation. The histone acetyltransferase (HAT) CBP is recruited by MLL via hydrophobic interactions in MLL aa 2829-2883.[11,12] MLL also interacts with the histone acetyltransferase MOF, an interaction that is important for transcriptional activation.[13]

Studies by several laboratories have shown that MLL is associated with mammalian homologs of proteins in the yeast Set1 methyltransferase complex. These include a core complex composed of hASH2, Rbb5, WDR5 and Dpy30.[13-15] This complex associates with the MLL SET domain and, via WDR5, targets MLL to sites of histone H3 lysine 4 dimethylation.[16] Importantly, the MLL complex also contains menin.[17] The exact function of menin is yet to be determined, although it is clear that menin is directly involved with transcriptional regulation. Menin has been reported to repress transcriptional activation by transcription factors JunD[18] and NF-κB.[19] However, our studies show that menin is involved in transcriptional activation. Menin interacts with the serine 5 phosphorylated form of RNA polymerase II (RNA pol II) and is required for transcription of target genes including the clustered *Hox* genes.[14,20] Menin interacts with MLL via a domain that is conserved in all leukemogenic MLL fusion proteins and apparently recruits MLL to target loci.

The best understood targets of MLL are the clustered homeobox or *Hox* genes, which are transcription factors that specify segment identity and cell fate during development. Previous studies showed that *Mll* (*Mll* = murine *MLL*) positively regulates *Hox* gene expression during development[21] because heterozygous *Mll* knockout mice showed posterior shifts in *Hox* gene expression. Moreover, *Mll* knockout mice are embryonic lethals in which patterns of *Hox* expression initiate normally, but are not maintained past embryonic day 9.5, when *Hox* expression drops to undetectable levels, indicating a pivotal role for MLL in maintenance of *Hox* gene expression.

Work done by our laboratory and others has also identified additional MLL/menin targets that are outside of the clustered *Hox* genes. Building on insights gained from growth data and microarray gene expression analysis on Mll and menin knockout fibroblasts, we determined that MLL directly regulated the expression of the cyclin-dependent kinase inhibitors *$p27^{Kip1}$* and *$p18^{Ink4c}$*.[22] We found that menin activates transcription via a mechanism involving recruitment of MLL to the *$p27^{Kip1}$* and *$p18^{Ink4c}$* promoters and coding regions. Loss of function of MLL or menin, either through ablation or expression of mutant alleles found in patients, results in down regulation of *$p27^{Kip1}$* and *$p18^{Ink4c}$* expression and deregulated cell growth. These findings were further extended by analyzing a series of pancreatic and parathyroid tumors from MEN1 patients. These studies confirmed a marked decrease in *$p27^{Kip1}$* in tumoral tissues. In aggregate, our data suggest that regulation of CDK inhibitor transcription by cooperative interaction between menin and MLL plays a central role in menin's activity as a tumor suppressor. These findings have been subsequently confirmed by other investigators.[23-25]

Role of Menin in Hematopoiesis

Normal hematopoiesis is markedly impaired in the absence of Mll and significant, but not overlapping, hematopoietic defects have now been identified in conditional menin knockout mice.

In the embryo, Mll is required to establish normal primitive and definitive hematopoiesis.[26-28] In addition, recent evidence shows that Mll is essential to support the homeostasis of adult hematopoietic progenitors.[29,30] In one of the two reports investigating this question, loss of Mll led to rapid and profound hematopoietic failure.[29] This was associated with an initial decrease in the quiescence of hematopoietic stem cells (HSCs), followed by HSC loss, as well as with downregulated expression of *Hoxa9*, *Hoxa7* and other clustered *Hox* genes. In the other report using an alternative approach to inactivate the *Mll* gene, steady-state hematopoiesis was less severely impaired, but Mll-deficient HSCs were markedly defective upon competitive transplantation into lethally irradiated hosts.[30] The basis for this difference in phenotypic severity is unclear, although it was likely related to differences in genetic strategies and may have resulted from incomplete elimination of Mll function in the mice with the less severe phenotype.[30] Altogether, this work identified key roles for Mll both in the establishment of hematopoiesis during embryonic development and its subsequent maintenance throughout life.

In view of these findings and because menin interacts with the Mll complex in leukemic cells, it was important to delineate the physiological impact of menin on normal hematopoiesis. Menin-deficient mice die during mid-gestation with multiple developmental defects.[31] Therefore, evaluating the role of menin in hematopoiesis required the use of a conditional *Men1* allele.[32] *Men1* inactivation in adult mice led to a modest decrease in the peripheral blood white cell count, as well as to a decreased ability of bone marrow progenitors to generate colonies in methylcellulose assays.[33] Unlike in Mll-deficient mice, loss of menin did not lead to overt hematopoietic failure. We have now investigated in detail the function of menin-deficient hematopoietic progenitors.[34] In the absence of hematopoietic stress, menin-deficient mice were able to maintain normal numbers of primitive hematopoietic progenitors containing hematopoietic stem cells, nonself-renewing multipotent progenitors and myeloerythroid progenitors. However, although common lymphoid progenitors were preserved, numbers of downstream B lineage progenitors were significantly decreased in the bone marrow, indicating that the lymphoid lineage is particularly sensitive to the loss of menin. In contrast to the mild defects observed during steady-state hematopoiesis, menin-deficient hematopoietic stem cells had a severely impaired repopulation potential in competitive transplantation assays and were also defective after drug-mediated chemoablation. Altogether, this discrepancy between a relatively well preserved steady-state hematopoiesis and profoundly abnormal HSC function after transplantation or chemoablation points to a specific role of menin in the adaptive response of HSCs to hematopoietic stress, a situation that involves the recruitment of quiescent HSCs into a burst of rapid proliferation.

The preservation of relatively normal hematopoiesis at steady-state in the absence of menin contrasts with the profound hematopoietic failure reported after loss of Mll, suggesting that menin may be absolutely required only for a subset of Mll's functions.[29,33,34] It remains to be determined if other proteins can substitute for menin to support Mll function in certain conditions, or if Mll can exert some of its effects totally independently of menin. An improved understanding of menin's precise role in the Mll complex during transcriptional regulation will be important to answer this question.

The relevant target genes of menin in the hematopoietic system remain to be identified. *Hoxa9* deficiency causes defects in HSC and lymphoid progenitor function that share characteristics with the defects observed after menin loss.[35,36] Therefore, it was tempting to speculate that reduced *Hoxa9* expression would account for at least some of the defects of menin-deficient progenitors. Intriguingly, we found that *Hoxa9* expression was normal in menin-deficient progenitor fractions containing HSCs.[34] These findings indicate that Mll and other regulatory inputs can maintain *Hoxa9* expression without menin in steady-state conditions. However, maintenance or induction of *Hoxa9* expression during a proliferative burst associated with hematopoietic stress may require menin-dependent epigenetic changes. Alternatively, the physiological role of menin in HSCs may not require *Hoxa9* at all and thus be dissociated from its role in supporting MLL fusion protein-mediated transformation.

Although it is clear that Hox genes are regulated by Mll and mediate many of its effects during transformation, the downstream genes that mediate Mll's effects in normal hematopoietic cells have also not been formally identified. Loss-of-function approaches have shown that the individual *Hox* genes examined so far do not support hematopoietic functions that are as prominent as the overall effect of Mll. Future work will have to establish if a combination of *Hox* genes regulated by Mll mediates its hematopoietic functions, or if Mll acts predominantly through non-*Hox* target genes.

The specific involvement of menin in HSC function during hematopoietic stress has several practical consequences. First, study of menin-deficient progenitors may give important insights in the regulation of the complex HSC response to situations of hematopoietic stress. This adaptive response is still poorly understood, yet it is functionally critical in many situations that are relevant to human health, such as hematopoietic recovery after chemotherapy or bone marrow transplantation. Of particular interest is the regulation of epigenetic changes that must occur in HSCs in this context and that may underlie the ability of HSCs to maintain expression of a functional stem cell program even while undergoing several rounds of rapid self-renewal divisions.

Role of Menin in Leukemogenesis

All the MLL fusion proteins examined to date upregulate expression of *Hoxa9* and *Meis1* and this appears to be pivotal for leukemogenesis. *Hox* genes including *Hoxa7* and *a9* and the *Hox* cofactor *Meis1* are normally only expressed in early Sca1+Lin-hematopoietic stem cells and then their expression is rapidly downregulated.[37-40] Although MLL is expressed throughout hematopoietic differentiation, normally *Hox* gene and *Meis1* expression is physiologically down modulated. In the presence of MLL fusion proteins, this mechanism is perturbed. In keeping with this, human leukemias with MLL rearrangements, either lymphoid or myeloid, consistently express *HOXA7*, *HOXA9* and *MEIS1*.[41-43] Experimental models provide strong evidence that upregulation of *Hox* genes, particularly *Hoxa9* and *Meis1*, accounts for MLL fusion protein leukemogenicity. *Hoxa7* and *Hoxa9* are consistently expressed in leukemias arising in BXH2 as a result of retroviral integration.[44,45] Notably, more than 95% of leukemias with *Hoxa7* and *a9* over expression show a second integration resulting in over expression of *Meis1*. Cotransduction of *Hoxa9* and *Meis1* immortalizes hematopoietic progenitors in vitro and rapidly accelerates leukemia development in transplanted mice.[37] These results are further supported by the inability of MLL fusion proteins to transform *Hoxa9* knockout bone marrow.[46]

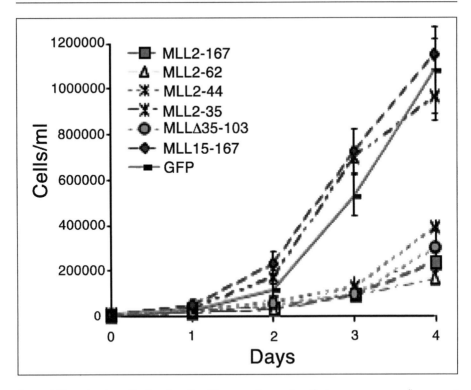

Figure 2. Transduction of leukemic cells with expression vectors that coexpress green fluorescent protein and peptides that inhibit the MLL-menin interaction inhibits growth. Growth curve analysis of cells purified by sorting for GFP expression. All peptides that inhibit MLL-menin interaction (MLL2-167, MLL2-62, MLL 2-44, MLLΔ35-103 inhibit growth while those that do not block the interaction (MLL2-35, MLL15-167, GFP) do not. Points show means of triplicate experiments, while bars show standard deviation. (Reproduced with permission from Caslini et al. Cancer Res 67:7275-83, 2007).

Menin is required for transcriptional activation and transformation by MLL fusion proteins. The protein binds to the N terminal 44 aa of MLL that are remote from the SET domain or MOF interaction domains. Previously we showed that menin interacts with the serine 5 phosphorylated form of RNA polymerase II[20] and in addition found that in fibroblasts menin appears to be important for recruitment of the MLL methyltransferase complex to target promoters.[10] It is likely that similar mechanisms are operative in hematopoietic cells. Deletion of the menin interaction domain from MLL fusion proteins results in complete loss of immortalizing ability. Furthermore we showed that dominant negative inhibitors of the MLL-menin interaction, which were derived from N-terminal MLL peptides inhibited the growth of MLL transformed cells (Fig.2). This is accompanied by down regulation of MLL targets including *Hox* genes and *Meis1*.[47]

Conclusion

Menin functions as an essential partner of MLL proteins within a large multiprotein complex with homology to the yeast Set1 methyltransferase complex. Although the precise biochemical mechanisms of menin's action remain to be fully investigated, it is clear that menin contributes to MLL-mediated Histone 3 Lysine 4 methylation at target gene loci. Menin regulates the homeostasis of normal hematopoietic progenitors. In addition, menin appears essential to mediate the transcriptional effects of MLL fusion proteins in leukemic cells, such as upregulation of *Hox* gene expression. As menin has only modest effects on hematopoietic stem cells in steady-state conditions, our findings suggest the existence of a therapeutic window to target the menin-MLL interaction in leukemia stem cells while sparing adjacent normal stem cells, at least in the absence of hematopoietic stress. These findings suggest that targeting the MLL-menin interaction is a promising target for leukemias with MLL rearrangements and possibly other leukemias with high level *Hox* and *Meis1* expression.

References

1. Felix CA. Secondary leukemias induced by topoisomerase-targeted drugs. Biochim Biophys Acta 1998; 1400:233-55.
2. Yokoyama A, Kitabayashi I, Ayton PM et al. Leukemia proto-oncoprotein MLL is proteolytically processed into 2 fragments with opposite transcriptional properties. Blood 2002; 100:3710-8.
3. Nakamura T, Mori T, Tada S et al. ALL-1 is a histone methyltransferase that assembles a supercomplex of proteins involved in transcriptional regulation. Mol Cell 2002; 10:1119-28.
4. Birke M, Schreiner S, Garcia-Cuellar MP et al. The MT domain of the proto-oncoprotein MLL binds to CpG-containing DNA and discriminates against methylation. Nucleic Acids Res 2002; 30:958-65.
5. Ayton PM, Chen EH, Cleary ML. Binding to nonmethylated CpG DNA is essential for target recognition, transactivation and myeloid transformation by an MLL oncoprotein. Mol Cell Bio 2004; 24:10470-8.
6. Schultz DC, Friedman JR, Rauscher FJ 3rd. Targeting histone deacetylase complexes via KRAB-zinc finger proteins: the PHD and bromodomains of KAP-1 form a cooperative unit that recruits a novel isoform of the Mi-2alpha subunit of NuRD. Genes Dev 2001; 15:428-43.
7. Fair K, Anderson M, Bulanova E et al. Protein interactions of the MLL PHD fingers modulate MLL target gene regulation in human cells. Mol Cell Biol 2001; 21:3589-97.
8. Filetici P, PO, Ballario P. The bromodomain: a chromatin browser? Front Biosci 2001; 6:866-76.
9. Milne TA, Briggs SD, Brock HW et al. MLL targets SET domain methyltransferase activity to hox gene promoters. Mol Cell 2002; 10:1107-17.
10. Milne TA, Dou Y, Martin ME et al. MLL associates specifically with a subset of transcriptionally active target genes. Proc Nat Acad Sci USA 2005; 102:14765-70.
11. Ernst P, Wang J, Huang M et al. MLL and CREB bind cooperatively to the nuclear coactivator CREB-binding protein. Mol Cell Biol 2001; 21:2249-58.
12. Petruk S, Sedkov Y, Smith S et al. Trithorax and dCBP acting in a complex to maintain expression of a homeotic gene. Science 2001; 294:1331-4.
13. Dou Y, Milne TA, Tackett AJ et al. Physical association and coordinate function of the H3 K4 methyltransferase MLL1 and the H4 K16 acetyltransferase MOF. Cell 2005; 121:873-85.
14. Yokoyama A, Wang Z, Wysocka J et al. Leukemia proto-oncoprotein MLL forms a SET1-like histone methyltransferase complex with menin to regulate hox gene expression. Mol Cell Biol 2004; 24:5639-49.
15. Milne TA, Hughes CM, Lloyd R et al. Menin and MLL cooperatively regulate expression of cyclin-dependent kinase inhibitors. Proc Nat Acad Sci USA 2005; 102:749-54.
16. Wysocka J, Swigut T, Milne TA et al. WDR5 associates with histone H3 methylated at K4 and is essential for H3 K4 methylation and vertebrate development. Cell 2005; 121:859-72.
17. Agarwal SK, Lee Burns A, Sukhodolets KE et al. Molecular pathology of the MEN1 gene. Ann N Y Acad Sci 2004; 1014:189-98.
18. Agarwal SK, Guru SC, Heppner C et al. Menin interacts with the AP1 transcription factor JunD and represses JunD-activated transcription. Cell 1999; 96:143-52.
19. Heppner C, Bilimoria KY, Agarwal SK et al. The tumor suppressor protein menin interacts with NF-kappaB proteins and inhibits NF-kappaB-mediated transactivation. Oncogene 2001; 20:4917-25.
20. Hughes CM, Rozenblatt-Rosen O, Milne TA et al. Menin associates with a trithorax family histone methyltransferase complex and with the hoxc8 locus. Mol Cell 2004; 13:587-97.
21. Yu BD, Hess JL, Horning SE et al. Altered hox expression and segmental identity in Mll-mutant mice. Nature 1995; 378:505-8.
22. Milne TA, Hughes CM, Lloyd R et al. Menin and MLL cooperatively regulate expression of cyclin-dependent kinase inhibitors. Proc Natl Acad Sci USA 2005; 102:749-54.
23. Karnik SK, Hughes CM, Gu X et al. Menin regulates pancreatic islet growth by promoting histone methylation and expression of genes encoding p27Kip1 and p18INK4c. Proc Nat Acad Sci USA 2005; 102:14659-64.
24. Scacheri PC, Davis S, Odom DT et al. Genome-wide Analysis of menin binding provides insights into MEN1 tumorigenesis. PLoS Genetics 2006; 2:e51.

25. Schnepp RW, Chen YX, Wang H et al. Mutation of tumor suppressor gene men1 acutely enhances proliferation of pancreatic islet cells. Cancer Res 2006; 66(11):5707-15.
26. Hess JL, Yu BD, Li B et al. Defects in yolk sac hematopoiesis in Mll-null embryos. Blood 1997; 90:1799-806.
27. Yagi H, Deguchi K, Aono A et al. Growth disturbance in fetal liver hematopoiesis of Mll-mutant mice. Blood 1998; 92:108-17.
28. Ernst P, Fisher JK, Avery W et al. Definitive hematopoiesis requires the mixed-lineage leukemia gene. Dev Cell 2004; 6:437-43.
29. Jude CD, Climer L, Xu D et al. Unique and independent roles for MLL in adult hematopoietic stem cells and progenitors. Cell Stem Cell 2007; 1:324-337.
30. McMahon KA, Hiew SYL, Hadjur S et al. Mll has a critical role in fetal and adult hematopoietic stem cell self-renewal. Cell Stem Cell 2007; 1:338-345.
31. Bertolino P, Radovanovic I, Casse H et al. Genetic ablation of the tumor suppressor menin causes lethality at mid-gestation with defects in multiple organs. Mech Dev 2003; 120:549-60.
32. Crabtree JS, Scacheri PC, Ward JM et al. Of mice and MEN1: Insulinomas in a conditional mouse knockout. Mol Cell Biol 2003; 23:6075-85.
33. Chen Y-X, Yan J, Keeshan K et al. The tumor suppressor menin regulates hematopoiesis and myeloid transformation by influencing hox gene expression. Proc Natl Acad Sci USA 2006; 103:1018-23.
34. Maillard I, Chen Y, Tubbs AT et al. Menin regulates the function of hematopoietic stem cells and lymphoid progenitors. Blood 2007; 110:379a.
35. Lawrence HJ, Helgason CD, Sauvageau G et al. Mice bearing a targeted interruption of the homeobox gene HOXA9 have defects in myeloid, erythroid and lymphoid hematopoiesis. Blood 1997; 89:1922-30.
36. Lawrence HJ, Christensen J, Fong S et al. Loss of expression of the hoxa-9 homeobox gene impairs the proliferation and repopulating ability of hematopoietic stem cells. Blood 2005; 106:3988-94.
37. Kroon E, Krosl J, Thorsteinsdottir U et al. Hoxa9 transforms primary bone marrow cells through specific collaboration with meis1a but not Pbx1b. EMBO J 1998; 17:3714-25.
38. Magli MC, Largman C, Lawrence HJ. Effects of HOX homeobox genes in blood cell differentiation. J Cell Physiol 1997; 173:168-77.
39. Lawrence HJ, Sauvageau G, Humphries RK et al. The role of HOX homeobox genes in normal and leukemic hematopoiesis. Stem Cells 1996; 14:281-91.
40. Pineault N, Helgason CD, Lawrence HJ et al. Differential expression of hox, meis1 and Pbx1 genes in primitive cells throughout murine hematopoietic ontogeny. Exp Hematol 2002; 30:49-57.
41. Armstrong SA, Staunton JE, Silverman LB et al. MLL translocations specify a distinct gene expression profile that distinguishes a unique leukemia. Nat Genet 2002; 30:41-7.
42. Rozovskaia T, Feinstein E, Mor O et al. Upregulation of Meis1 and HoxA9 in acute lymphocytic leukemias with the t(4:11) abnormality. Oncogene 2001; 20:874-8.
43. Yeoh EJ, Ross ME, Shurtleff SA et al. Classification, subtype discovery and prediction of outcome in pediatric acute lymphoblastic leukemia by gene expression profiling. Cancer Cell 2002; 1:133-43.
44. Nakamura T, Largaespada DA, Shaughnessy JD Jr et al. Cooperative activation of hoxa and Pbx1-related genes in murine myeloid leukaemias. Nat Genet 1996; 12:149-53.
45. Moskow JJ, Bullrich F, Huebner K et al. Meis1, a PBX1-related homeobox gene involved in myeloid leukemia in BXH-2 mice. Mol Cell Biol 1995; 15:5434-43.
46. Ayton PM, Cleary ML. Molecular mechanisms of leukemogenesis mediated by MLL fusion proteins. Oncogene 2001; 20:5695-707.
47. Caslini C, Yang Z, El-Osta M et al. Interaction of MLL amino terminal sequences with menin is required for transformation. Cancer Res 2007; 67:7275-83.

CHAPTER 6

Role of Menin in Bone Development

Hiroshi Kaji,* Lucie Canaff and Geoffrey N. Hendy

Abstract

Menin function is related to transcriptional regulation and cell cycle control and it physically and functionally interacts with osteotropic transcription factors, such as Smad1/5, Smad3, Runx2 and JunD. Menin promotes the commitment of pluripotent mesenchymal stem cells to the osteoblast lineage, mediated by interactions between menin and the BMP signaling molecules, Smad1/5, or Runx2. On the other hand, in mature osteoblasts the interaction of menin and the TGF-β/Smad3 pathway counteracts the BMP-2/Smad1/5- and Runx2-induced transcriptional activities leading to inhibition of late stage osteoblast differentiation. Moreover, menin suppresses osteoblast maturation partly by inhibiting the differentiation actions of JunD. In conclusion, menin plays an important role in osteoblastogenesis and osteoblast differentiation.

Introduction

Osteoblasts, chondrocytes, adipocytes and myoblasts are derived from common precursor cells—multipotential mesenchymal stem cells derived from mesoderm. Several hormones, growth factors and cytokines regulate, systemically and locally, the commitment of the stem cells into osteoblastic cells and their subsequent maturation. Moreover, transcriptional regulators, such as Runx2 and AP-1 family members, are critical for osteoblast differentiation.[1,2] Here, we review our studies that show the interaction of menin with the BMP-2, TGF-β and AP-1 pathways in osteoblastogenesis and differentiation.

Menin is expressed in all mouse tissues examined and these include bone.[3] Homozygous *MenI* inactivation in mice is embryonic lethal at 12 days and some fetuses exhibit clear defects in cranial and facial development, whereas the heterozygous phenotype is strikingly similar to that of MEN1 in humans, with endocrine tumors developing later in life.[4,5] Since cranial bones are formed by intramembranous ossification, the findings suggested that menin might play a role in bone formation. Our studies provide evidence of roles for menin in the commitment of multipotential mesenchymal stem cells to the osteoblast lineage and the later differentiation of osteoblasts.

Menin and TGF-β Signaling

We demonstrated initially that menin interacts with the transforming growth factor (TGF)-β signaling molecule, Smad3, in the rat anterior pituitary GH4C1 cell line.[6] Menin inactivation by antisense RNA antagonizes TGF-β-mediated cell proliferation inhibition. Moreover, menin inactivation with antisense RNA suppresses TGF-β-induced and Smad3-induced transcriptional activity by inhibiting Smad3/4-DNA binding at specific transcriptional regulatory sites. Given the important function of TGF-β acting to negatively regulate cell proliferation in an autocrine or paracrine fashion, the loss of menin/Smad3 interactions has clear implications for cell growth dysregulation leading to tumorigenesis. Tumor suppressors often function as checkpoints, ensuring

*Corresponding Author: Hiroshi Kaji—Division of Diabetes, Metabolism and Endocrinology, Department of Internal Medicine, Kobe University Graduate School of Medicine. 7-5-2 Kusunoki-cho, Chuo-ku, Kobe 650-0017, Japan. Email: hiroshik@med.kobe-u.ac.jp

SuperMEN1: Pituitary, Parathyroid and Pancreas, edited by Katalin Balogh and Attila Patocs.
©2009 Landes Bioscience and Springer Science+Business Media.

that the cell cycle is arrested when the cell is exposed to inhibitory growth factors.[7] Blockage of TGF-β signaling may disrupt the delicately balanced cellular steady state, pushing the cell toward inappropriate growth that ultimately results in tumor formation. Indeed, several reports have described inactivating mutations in genes encoding proteins known to be essential for the TGF-β signaling pathways including genes for Smad2 and Smad4, in cancers of pancreas, biliary tract, colon, lung, head, neck and hepatocellular carcinoma.[8]

Menin and AP-1 Signaling

The AP-1 transcription factor family consists of Jun members (c-Jun, JunB and JunD) and Fos members (c-Fos, FosB, Fra1 and Fra2). Among them, only JunD has been demonstrated to directly interact with menin.[9,10] Menin represses JunD-activated transcription by association with an mSin3A-histone deacetylase complex.[9,11] Only the full-length isoform of JunD binds to menin.[12] We found that JunD expression was higher in proliferating than resting anterior pituitary GH4C1 cells,[13] although, in general, the expression of JunD is constitutive and is less affected by growth factor and external stimuli.[14] In the anterior pituitary cells the expression of JunD paralleled that of menin. Data have been presented to support the hypothesis that the character of JunD changes depending upon whether it is bound or not bound to menin. Thus, JunD changed from growth suppressor to growth promoter when its binding to menin was prevented by it being mutated or by being in a menin-null genetic background.[15] In COS cells, menin augmented c-Jun-mediated transactivation by an unknown mechanism.[16]

BMP, TGF-β and AP-1 Signaling in the Osteoblast

BMP-2 has the ability to induce ectopic bone and cartilage formation in extraskeletal tissues in vivo.[2] BMP-2 plays critical roles in bone formation and bone cell differentiation and it induces mesenchymal stem cells to differentiate into osteoblasts.[17]

TGF-β is most abundant in the bone matrix. It is stored in an inactive form then released and activated in the bone microenvironment.[18] TGF-β generally inhibits osteoblast differentiation. However, we found that Smad3, normally considered to mediate TGF-β signaling, enhanced alkaline phosphatase (ALP) activity, mineralization and the level of bone matrix proteins such as Type 1 collagen (Col1), suggesting that Smad3 in some novel fashion plays an important role in osteoblastic bone formation.[19] This discordance between TGF-β and Smad3 actions was likely due to the activation of ERK1/2 and JNK by TGF-β negatively regulating the Smad3-induced ALP activity and mineralization in osteoblasts.[20]

BMP-2 and TGF-β act through cell-surface complexes of Type I and Type II transmembrane serine/threonine kinase receptors.[18] In the presence of ligand, the receptors associate and the Type II receptor phosphorylates the Type I receptor, which propagates a signal by phosphorylating receptor-regulated Smads at their carboxyl-termini and activating them. BMP receptors activate Smad1, Smad5 and Smad8 and TGF-β receptors activate Smad2 and Smad3. Activated Smads form stable complexes with a common Smad, Smad4 and the complex translocates to the nucleus and regulates transcriptional responses initiated by the BMP or TGF-β (Fig. 1).

Runx2 (also known as Cbfa1) is a master gene for the generation of fully functional osteoblasts. Mice homozygous null for Runx2 have complete lack of bone formation with arrest of osteoblast differentiation.[21] Runx2 is a common target of BMP-2 and TGF-β (Fig. 2) and cooperation between Runx2 and the BMP-2 signaling molecule, Smad5, induces osteoblast-specific gene expression in mesenchymal stem cells. Runx2 induces bone matrix genes, including Col1, osteopontin (OPN), bone sialoprotein, osteocalcin (OCN) and fibronectin.[2] On the other hand, Runx2 inhibits maturation. Its positive effects on differentiation are restricted to early stage osteoblast development.

Osterix (Osx) is a zinc finger transcription factor specific to the osteoblast lineage and critical for formation of immature osteoblasts from preosteoblasts and bone differentiation.[22] Runx2 positively regulates the activity of the Osx gene (Fig. 2)[23] whereas the tumor suppressor, p53, represses both Runx2 and Osx transcription and osteoblastogenesis.[24,25] Thus far, interactions between Osx and menin are not known.

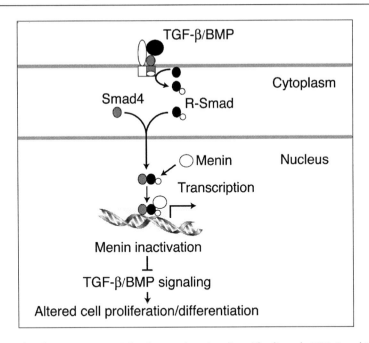

Figure 1. Role of menin in TGF-β family member signaling. The ligands TGF-β and BMP signal through Type I and Type II serine/threonine kinase receptors. Receptor-activated Smads (R-Smad: Smad2 and 3, TGF-β; Smad1, 5 and 8; BMP) associate with the common Smad (Smad4) and translocate to the nucleus. In the nucleus, menin is bound to the R-Smad and facilitates transcription by the Smad complex. If menin is inactivated, TGF-β ligand signaling is blocked leading to altered cell proliferation and/or differentiation.

Wnts are secreted glycoproteins that mediate local cellular interactions in development. Canonical Wnt signaling (via β-catenin) critically controls osteoblastogenesis.[26] Interaction of Wnt proteins with the cell-surface frizzled (FRP) receptors and low density lipoprotein-related protein (LRP)-5 and -6 coreceptors results in glycogen synthase kinase (GSK)-3β-mediated β-catenin phosphorylation resulting in β-catenin accumulation and translocation to the nucleus and upregulation of gene transcription by interaction with high-mobility group box transcriptional factors of the lymphoid enhancer-binding factor (Lef)/T cell factor (Tcf) family (Fig. 2). We have shown that the canonical Wnt-β-catenin pathway is involved in the osteoblast anti-apoptotic actions of parathyroid hormone (PTH).[27] This occurs in part via Smad3 although, as yet, interactions between menin and the Wnt pathway have not been explored.

With respect to actions of AP-1 family members in bone, overexpression of c-Fos in mice causes osteosarcoma and homozygous c-fos-deficient mice develop osteopetrosis with a lack of osteoclasts.[28,29] Overexpression of either Fra-1 or DeltafosB increases bone formation causing osteosclerosis in transgenic mice.[30,31] Loss of JunB in mice results in reduced bone formation and severe bone turnover osteopenia mainly due to a cell-autonomous osteoblast and osteoclast differentiation defect.[32] Thus, AP-1 signaling is also important for bone remodeling (Fig. 2). As for JunD, its expression in osteoblasts is modulated by external stimuli, either mechanical strain or basic fibroblast growth factor.[33,34] Overexpression of JunD and Fra2 represses Runx2-induced collagenase-3 gene promoter activity in differentiated osteoblasts, suggesting that Runx2/AP-1 interaction regulates this matrix metalloprotease important for skeletal development and normal and pathological remodeling of bone.[35] Moreover, expression of OCN is upregulated by overexpression of JunD and Fra2 in rat osteoblasts.[36] Diminished AP-1 activity, especially JunD and the

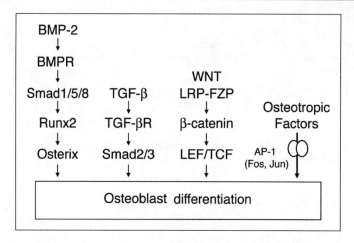

Figure 2. Factors that modulate osteoblastogenesis and osteoblast differentiation.

resultant decline in the expression of interleukin-11 in bone marrow stromal cells plays a role in the impaired bone formation of a strain of senescence-accelerated mice.[37]

Role of Menin in Early Stage Osteoblast Differentiation

We examined changes in menin expression levels in mouse uncommitted mesenchymal cell-lines (10T1/2, ST2 and PA6 cells) and mouse osteoblastic MC3T3-E1 cells during the progression of differentiation.[38] In confluent 10T1/2 cells, at 7 days of culture, menin was expressed and the levels increased with BMP-2 treatment by 14 days but had decreased somewhat at 21 days (Fig. 3a). While Col1 and Runx2 remained constant, OCN expression increased throughout differentiation, as did ALP production (Fig. 3b). Treatment of confluent 10T1/2 cells for 12 h with menin antisense oligonucleotides (AS-oligo) suppressed the expression of endogenous menin (Fig. 3c). When these cells were stimulated with BMP-2, the normal increases in Col1 (Fig. 3d) and ALP activity (Fig. 3e) and other osteoblast markers (Runx2 and OCN) were antagonized. However, menin inactivation did not affect adipogenic markers (Fig. 3f) or chondrocytic markers induced by BMP-2. These findings indicate that menin is a crucial factor in the commitment of multipotential mesenchymal cells to the osteoblast lineage. In the osteoblastic MC3T3-E1 cells, menin expression was well expressed at days 7 and 14 and had declined at day 21, a period of mineralization (Fig. 3g). Although AS-oligo treatment suppressed endogenous menin expression (Fig. 3h) in the MC3T3-E1 cells, it did not affect BMP-2-induced ALP activity (Fig. 3i) or expression of the other osteoblastic markers.

Menin was co-immunoprecipitated with Smad1/5 in ST2 mesenchymal cells and MC3T3-E1 cells and inactivation of menin antagonized the BMP-2-induced transcriptional activity of Smad1/5 in ST2 cells, but not MC3T3-E1 cells.[39] Moreover, menin was co-immunoprecipitated with Runx2 and menin inactivation with AS-oligo antagonized Runx2 transcriptional activity and the ability of Runx2 to stimulate ALP activity only in ST2 cells but not in MC3T3-E1 cells. These findings indicate that menin interacts physically and functionally with Runx2 in uncommitted mesenchymal stem cells, but not well-differentiated osteoblasts.

Menin and TGF-β Pathway in Osteoblast Differentiation

In osteoblastic MC3T3-E1 cells, TGF-β and Smad3 negatively regulate Runx2 transcriptional activity (Fig. 4a). Menin and Smad3 co-immunoprecipitated (Fig. 4b) and the co-expression of menin and Smad3 antagonized BMP-2-induced transcriptional activity of Smad1/5 and Runx2. Menin inactivation with AS-oligo treatment antagonized the Smad3-driven luciferase activity of the Smad3-response element-containing 3TP-lux reporter construct (Fig. 4c). Thus, the TGF-β/

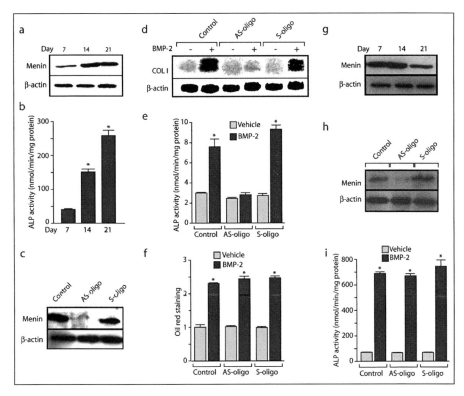

Figure 3. Role of menin in osteoblastogenesis and early osteoblast differentiation. a) Menin expression increases over time in 10T1/2 mesenchymal stem cells treated with BMP-2. b) The production of alkaline phosphatase (ALP) increases with BMP-2 in a time-dependent manner in 10T1/2 cells. c) Antisense menin (AS-oligo) treatment inhibits menin expression in confluent (7 day) cultures of 10T1/2 cells, whereas sense menin (S-oligo) treatment is without effect. AS-oligo treatment of 10T1/2 cells specifically antagonizes the BMP-2-induced expression of, d) collagen Type I (COLI) and e) ALP activity, but has no effect on f) adipocyte production as indicated by Oil red staining. g) Menin expression is high initially but declines as osteoblastic MC3T3-E1 cells differentiate. h) AS-oligo treatment of MC3T3-E1 cells specifically inhibits menin expression but i) in contrast to the effect in 10T1/2 cells has no effect on BMP-2-induced ALP activity. For experimental details see reference 38.

Smad3 pathway negatively regulates BMP/Smad1/5- and Runx2-induced transcriptional activities leading to inhibition of late stage osteoblast differentiation.[39]

With respect to differentiation of mature osteoblasts, menin inactivation in MC3T3-E1 cells with menin antisense oligonucleotides affected neither BMP-2-stimulated ALP activity nor the expression of Runx2 and OCN.[38] Inactivation of menin by stable transfection of an antisense menin cDNA in MC3T3-E1 cells increased ALP activity, mineralization and the expression of Col1 and OCN.

In mesenchymal stem cells, BMP-2 activates Smad1/5 and menin interacts with Smad1/5 physically and functionally, resulting in increased expression and activation of Runx2 (Fig. 5a). Moreover, menin interacts with Runx2 physically and functionally, promoting the commitment of the multipotential mesenchymal stem cells to the osteoblast lineage. After commitment to the osteoblast lineage, the role of menin changes and it no longer exerts a positive influence on the BMP-2/Smad1/5/Runx2 pathway. Menin is then important for the effects of TGF-β-activated

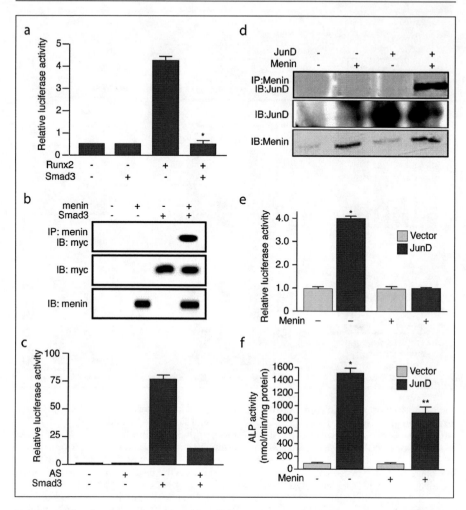

Figure 4. Role of menin in later osteoblast differentiation. In osteoblastic MC3T3-E1 cells, a) Smad3 negatively regulates Runx2 transcriptional activity, b) Menin and Smad3 co-immunoprecipitate and c) menin inactivation with AS-oligo antagonizes the Smad3-driven luciferase activity of the Smad3-response-element containing 3TP-lux reporter construct. For experimental details see reference 39. In osteoblastic MC3T3-E1 cells, JunD enhances differentiation and d) menin and JunD co-immunoprecipitate. e) JunD overexpression increases the activity of an AP-1 promoter-luciferase reporter construct and the activity is reduced by cotransfection of menin. f) Menin overexpression inhibits the alkaline phosphatase activity (ALP) induced by JunD. For experimental details see reference 40.

Smad3 in inhibiting the BMP-2/Runx2 cascade (Fig. 5a). BMP- and TGF-β-signaling pathways may cross-talk via menin.

Menin and JunD in the Osteoblast

JunD is expressed in osteoblasts where it interacts with menin.[40] JunD expression increases gradually during osteoblast differentiation. Stable expression of JunD enhanced expression of Runx2, Col1, OCN, ALP and mineralization in MC3T3-E1 cells, indicating that JunD enhances osteoblast differentiation. In MC3T3-E1 cells in which menin expression was reduced by stable

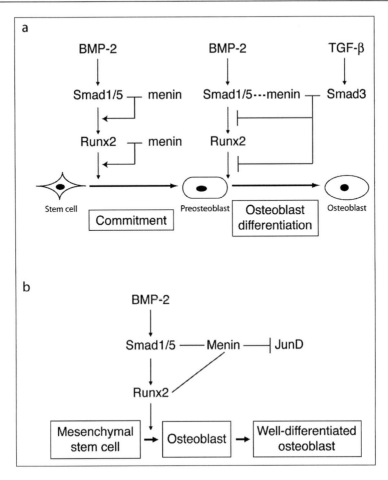

Figure 5. a) Role of menin in osteoblastogenesis and osteoblast differentiation. b) Role of JunD-menin interaction in osteoblast maturation.

antisense menin cDNA transfection, JunD levels were increased. When JunD and menin sense constructs were cotransfected in MC3T3-E1 cells, the proteins co-immunoprecipitated (Fig. 4d). JunD overexpression increased the activity of an AP-1 promoter-luciferase reporter construct and this activity was reduced by cotransfection of menin cDNA (Fig. 4e). These findings indicate that JunD and menin interact both physically and functionally in osteoblasts. Furthermore, menin overexpression inhibited the ALP activity induced by JunD (Fig. 4f). Taken together, menin suppresses osteoblast maturation, in part, by inhibiting the differentiation actions of JunD (Fig. 5b).

Conclusion

Menin plays significant roles in osteoblast differentiation through its complex interaction with BMP-Runx2, TGF-β and JunD pathways. Menin interacts with other bone-related factors, such as retinoblastoma protein, estrogen receptor, Hox and heat shock proteins, insulin-like growth factor-binding protein-2 and telomerase.[41-48] Moreover, menin induces apoptosis in murine embryonic fibroblasts and could potentially be involved in apoptosis of osteoblasts.[49] In summary, menin may subserve pleiotropic actions in bone. Since MenI-homozygous null mice are embryonic lethal, knowledge of the in vivo functions of menin specifically in bone is lacking. Recent studies revealed that tissue-specific MenI deletion in Pax3- or Wnt1- expressing neural crest cells leads to perinatal death, cleft palate and other cranial bone defects following defective bone mineralization. Moreover, the absence of menin also resulted in defective rib formation.[50] The generation and analysis of mouse strains in which MenI is specifically overexpressed in or deleted from bone cells will be necessary to further understand the role of menin in bone development.

Acknowledgements

Work from our laboratories has been supported by the Kanzawa Medical Research Foundation (to H.K.) and the Ministry of Science, Education and Culture of Japan (Grant-in-aid 15590977 to H.K.) and the Canadian Institutes of Health Research (CIHR) (Grant MOP-9315 to G.N.H.).

References

1. Karsenty G, Wagner EF. Reaching a genetic and molecular understanding of skeletal development. Dev Cell 2002; 2:389-406.
2. Yamaguchi A, Komori T, Suda T. Regulation of osteoblast differentiation mediated by bone morphogenetic proteins, hedgehogs and Cbfa1. Endocr Rev 2000; 21:393-411.
3. Stewart C, Parente F, Piehl F et al. Characterization of the mouse Men1 gene and its expression during development. Oncogene 1998; 17:2485-93.
4. Crabtree JS, Scacheri PC, Ward JM et al. A mouse model of multiple endocrine neoplasia type 1, develops multiple endocrine tumors. Proc Natl Acad Sci USA 2001; 98:1118-23.
5. Bertolino P, Tong WM, Galendo D et al. Heterozygous men1 mutant mice develop a range of endocrine tumors mimicking multiple endocrine neoplasia type 1. Mol Endocrinol 2003; 17:1880-92.
6. Kaji H, Canaff L, Lebrun JJ et al. Inactivation of menin, a Smad3-interacting protein, blocks transforming growth factor type β signaling. Proc Natl Acad Sci USA 2001; 98:3837-42.
7. Hu PP, Datto MB, Wang XF. Molecular mechanisms of transforming factor-β signaling. Endocr Rev 1998; 19:349-63.
8. Yakicier MC, Irmak MB, Romano A et al. Smad2 and Smad4 gene mutations in hepatocellular carcinoma. Oncogene 1999; 18:4879-83.
9. Agarwal SK, Guru SC, Heppner C et al. Menin interacts with the AP1 transcription factor JunD and represses JunD-activated transcription. Cell 1999; 96:143-52.
10. Gobl AE, Berg M, Lopez-Egido LR et al. Menin represses JunD-activated transcription by a histone deacetylase-dependent mechanism. Biochim Biophys Acta 1999; 1447:51-6.
11. Kim H, Lee JE, Cho EJ et al. Menin, a tumor suppressor, represses JunD-mediated transcriptional activity by association with an mSin3A-histone deacetylase complex. Cancer Res 2003; 63:6135-9.
12. Yazgan O, Pfarr CM. Differential binding of the menin tumor suppressor protein to JunD isoforms. Cancer Res 2001; 61:916-20.
13. Kaji H, Canaff L, Goltzman D et al. Cell cycle regulation of menin expression. Cancer Res 1999; 59:5097-101.
14. Pfarr CM, Mechta F, Spyrou G et al. Mouse JunD negatively regulates fibroblast growth and antagonizes transformation by ras. Cell 1994; 76:747-60.
15. Agarwal SK, Novotny EA, Crabtree JS et al. Transcriptional factor JunD, deprived of menin, switches from growth suppressor to growth promoter. Proc Natl Acad Sci USA 2003; 100:10770-5.
16. Ikeo Y, Yumita W, Sakurai A et al. JunD-menin interaction regulates c-Jun-mediated AP-1 transactivation. Endocr J 2004; 51:333-42.
17. Wozney JM, Rosen V, Celeste AJ et al. Novel regulators of bone formation: Molecular clones and activities. Science 1988; 242:1528-34.
18. Janssens K, ten Dijke P, Janssens S et al. Transforming growth factor-β1 to the bone. Endocr Rev 2005; 26:743-74.
19. Sowa H, Kaji H, Yamaguchi T et al. Smad3 promotes alkaline phosphatase activity and mineralization of osteoblastic MC3T3-E1 cells. J Bone Miner Res 2002; 17:1190-9.
20. Sowa H, Kaji H, Yamaguchi T et al. Activations of ERK1/2 and JNK by transforming growth factor β negatively regulate Smad3-induced alkaline phosphatase activity and mineralization in mouse osteoblastic cells. J Biol Chem 2002; 277:36024-31.
21. Komori T, Yagi H, Nomura S et al. Targeted disruption of Cbfa1 results in a complete lack of bone formation owing to maturational arrest of osteoblasts. Cell 1997; 89:755-64.
22. Nakashima K, Zhou X, Kunkel G et al. The novel zinc finger-containing transcription factor osterix is required for osteoblast differentiation and bone formation. Cell 2002; 108:17-29.
23. Nashio Y, Dong Y, Paris M et al. Runx2-mediated regulation of the zinc finger Osterix/Sp7 gene. Gene 2006; 372:62-70.

24. Wang X, Kua HY, Hu Y et al. p53 functions as a negative regulator of osteoblastogenesis, osteoblast-dependent osteoclastogenesis and bone remodeling. J Cell Biol 2006; 172:115-125.
25. Lengner CJ, Steinman HA, Gagnon J et al. Osteoblast differentiation and skeletal development are regulated by Mdm2-p53 signaling. J Cell Biol 2006; 172:909-921.
26. Bodine PVN, Komm BS. Wnt signaling and osteoblastogenesis. Rev Endocr Metab Discord 2006; 7:33-39.
27. Tobimatsu T, Kaji H, Sowa H et al. Parathyroid hormone increases β-catenin levels through Smad3 in mouse osteoblastic cells. Endocrinol 2006: 147:2583-2590.
28. Grigoriadis A, Schellander K, Wang ZQ et al. Osteoblasts are target cells for transformation in c-fos transgenic mice. J Cell Biol 1993; 122:685-701.
29. Castellazzi M, Spyrou G, La Vista N et al. Overexpression of c-jun, junB, or junD affects cell growth differently. Proc Natl Acad Sci USA 1991; 88:8890-4.
30. Jochum W, David JP, Elliott C et al. Increased bone formation and osteosclerosis in mice overexpressing the transcription factor Fra-1. Nat Med 2000; 6:980-4.
31. Sabatakos G, Sims NA, Chen J et al. Overexpression of DeltaFosB transcription factor(s) increases bone formation and inhibits adipogenesis. Nat Med 2000; 6:985-90.
32. Kenner L, Hoebertz A, Beil T et al. Mice lacking JunB are osteopenic due to cell-autonomous osteoblast and osteoclast defects. J Cell Biol 2004; 164:613-23.
33. Granet C, Vico AG, Alexandre C et al. MAP and src kinases control the induction of AP-1 members in response to changes in mechanical environment in osteoblastic cells. Cell Signal 2002; 14:679-688.
34. Varghese S, Rydziel S, Canalis E. Basic fibroblast growth factor stimulates collagenase-3 promoter activity in osteoblasts through an activator protein-1-binding site. Endocrinology 2000; 141:2185-91.
35. Winchester SK, Selvamurugan N, D'Alonzo RC et al. Developmental regulation of collagenase-3 mRNA in normal, differentiating osteoblasts through the activator protein-1 and the runt domain binding sites J Biol Chem 2000; 275:23310-8.
36. McCabe LR, Banerjee C, Knudu R et al. Developmental expression and activities of specific fos and jun proteins are functionally related to osteoblast maturation: role of Fra-2 and JunD during differentiation. Endocrinology 1996; 137:4398-408.
37. Tohjima E, Inoue D, Yamamoto N et al. Decreased AP-1 activity and interleukin-11 expression by bone marrow stromal cells may be associated with impaired bone formation in aged mice. J Bone Miner Res 2003; 18:1461-70.
38. Sowa H, Kaji H, Canaff L et al. Inactivation of menin, the product of the multiple endocrine neoplasia type 1 gene, inhibits the commitment of multipotential mesenchymal stem cells into the osteoblast lineage. J Biol Chem 2003; 278:21058-69.
39. Sowa H, Kaji H, Hendy GN et al. Menin is required for bone morphogenetic protein 2- and transforming growth factor β-regulated osteoblastic differentiation through interaction with Smads and Runx2. J Biol Chem 2004; 279:40267-75.
40. Naito J, Kaji H, Sowa H et al. Menin suppresses osteoblast differentiation by antagonizing the AP-1 factor, JunD. J Biol Chem 2005; 280:4785-4791.
41. Loffler KA, Biondi CA, Gartside MG et al. Lack of augmentation of tumor spectrum or sensitivity in dual heterozygous Men1 and Rb1 knockout mice. Oncogene 2007; 26:4009-17.
42. Dreijerink KMA, Mulder KW, Winkler GS et al. Menin links estrogen receptor activation to histone H3K4 trimethylation. Cancer Res 2006; 66:4929-35.
43. Chen YX, Yan J, Keeshan K et al. The tumor suppressor menin regulates hematopoiesis and myeloid transformation by influencing Hox gene expression. Proc Natl Acad Sci USA 2006; 103:1018-23.
44. Hughes CM, Rosenblatt-Rosen O, Milne TA et al. Menin associates with a trithorax family histone methyltransferase complex and with the Hoxc8 locus. Mol Cell 2004; 13:587-97.
45. Yokoyama A, Wang Z, Wysocka J et al. Leukemia proto-oncoprotein MLL forms a SET1-like histone methyltransferase complex with menin to regulate Hox gene expression. Mol Cell Biol 2004; 24:5639-49.
46. Papaconstantinou M, Wu Y, Pretorius HK et al. Menin is a regulator of the stress response in drosophila melanogaster. Mol Cell Biol 2005; 25:9960-72.
47. Ping LA, Schnepp RW, Petersen CD et al. Tumor suppressor menin regulates expression of insulin-like growth factor binding protein 2. Endocrinology 2004; 145:3443-50.
48. Lin SY, Elledge SJ. Multiple tumor suppressor pathways negatively regulate telomerase. Cell 2003; 113:881-9.
49. Schnepp RW, Mao H, Sykes SM et al. Menin induces apoptosis in murine embryonic fibroblasts. J Biol Chem 2004; 279:10685-91.
50. Engleka KA, Wu M, Zhang M et al. Menin is required in cranial neural crest for palatogenesis and perinatal viability. Dev Biol 2007; 311:524-37.

CHAPTER 7

Activin, TGF-β and Menin in Pituitary Tumorigenesis

Jean-Jacques Lebrun*

Abstract

Pituitary adenomas are common monoclonal neoplasms accounting for approximately one-fifth of primary intracranial tumors. Prolactin-secreting pituitary adenomas (prolactinomas) are the most common form of pituitary tumors in humans. They are associated with excessive release of the hormone prolactin and increased tumor growth, giving rise to severe endocrine disorders and serious clinical concerns for the patients. Recent studies indicated that the activin/TGF-β family of growth factors plays a prominent role in regulating pituitary tumor growth and prolactin secretion from anterior pituitary lactotrope cells. Furthermore, these studies highlighted the tumor suppressor menin and the protein Smads as central regulators of these biological processes in the pituitary. Alterations in the activin/TGF-β downstream signaling pathways are critical steps towards tumor formation and progression. This chapter will review the role and intracellular molecular mechanisms of action by which activin, TGF-β, Smads and menin act in concert to prevent pituitary tumor cell growth and control hormonal synthesis by the anterior pituitary.

Introduction

The pituitary gland is the primary site of the synthesis, storage and release of hormones that play a predominant role within the entire human body and thus careful regulation of these hormonal levels is essential to maintain homeostasis.[1] Pituitary tumors account for 15% to 20% of clinically diagnosed intracranial tumors.[1] Although considered as histologically benign, pituitary tumors can cause significant morbidity, because of the excessive pituitary hormonal secretion, critical location and expanding size. Despite a critical improvement in recent technologies such as imaging and surgical endoscopy, the removal of pituitary tumors largely depends on the expertise of the surgeon. The most common type of pituitary adenomas are prolactinomas, tumors of the anterior pituitary prolactin-secreting lactotrope cells.[2] Patients with prolactinomas usually present amenorrhea, have infertility issues associated with galactorrhea in females and impotence in males infertility.[1-3] Prolactinomas often develop sporadically as a monoclonal proliferation but the molecular mechanisms underlying the formation of these tumors remain largely unknown. Besides surgical removal of the tumor, medical therapy for prolactinomas are effective in many cases but usually necessitates long-term treatment with dopamine agonists to normalize prolactin levels. Recent work from our laboratory shed light on the mechanisms by which activin and TGF-β regulate prolactin levels and cell growth arrest in lactotrope cells, through the Smad pathway and the tumor suppressor menin.[4,5] Such understanding of the signaling pathways that regulate pituitary

*Jean-Jacques Lebrun—Hormones and Cancer Research Unit, Department of Medicine, Royal Victoria Hospital, McGill University, 687 Pine Avenue West, H3A 1A1, Montreal, Quebec, Canada. Email: jj.lebrun@mcgill.ca

SuperMEN1: Pituitary, Parathyroid and Pancreas, edited by Katalin Balogh and Attila Patocs. ©2009 Landes Bioscience and Springer Science+Business Media.

hormonal production and cell growth may prove helpful to open new avenues for future therapies to combat human pituitary adenomas.

Genetics of Pituitary Adenomas: Role of MEN1 Mutations

Genetic alterations play an important role in the genesis and progression of pituitary adenomas. These alterations mainly occur in genes coding for tumor suppressors, oncogenes and transcription factors. The initiating event in pituitary adenoma development is primarily due to mutation in the stimulatory guanine nucleotide-binding protein (*gsp*) and *MEN1* genes. The progression event in pituitary adenomas occurs later in the development of the tumors and usually result from mutations in p53, ras, retinoblastoma, metastasis suppressor nm23 and c-myc genes.[6]

Mutations in the *gsp* and *MEN1* genes are critical to the initiation of pituitary tumors. Gsp is an oncogene that leads to increased cAMP production, GH hypersecretion and cell cycle progression.[7-9] The tumor suppressor gene *MEN1* is associated with MEN-1 syndrome and characterized by the occurrence of anterior pituitary, parathyroid and pancreatic islet tumors.[10] Even though *MEN1* gene mutation is usually associated with familial pituitary tumors,[10] mutation in the *MEN1* gene have also been detected in sporadic pituitary tumors.[11,12] These include as 28% of ACTH secreting adenomas,[13] 20% of the nonfunctional adenomas,[13] 15-30% of somatotroph adenomas[13,14] and 12-14% of prolactinomas.[13,15] Loss of heterozygosity (LOH) at the menin locus was also reported in anterior pituitary tumors.[16,17] The familial syndrome MEN1 behaves as an autosomal dominant trait with reduced penetrance. Germ-line mutation on chromosome 11q13 that encodes the tumor suppressor menin, is unmasked by a second somatic hit on the remaining allele. The human MEN1 syndrome phenotype is well reflected by the MEN1 heterozygous transgenic mice model which develops tumors with LOH of the wild type chromosome, including 26% of these within the pituitary by the age of 16 months.[18] Somatic mutations of the *MEN1* gene are not significantly causative in the tumorigenesis of non-MEN1-linked sporadic pituitary adenomas.[19,20] Indeed, out of 35 sporadic pituitary adenomas of various secretory phenotypes used in one study, Poncin et al found that only one tumor out of the cohort was homozygote for a mutation in the close proximity of the MEN1 gene promoter.[20] A more recent study, performed in a series of tissue samples from 68 sporadic nonMEN1 pituitary tumor patients further confirmed this and found only one case to show detectable menin expression, as measured by immunohistochemistry and immunofluorescence.[21] As mentioned above, pituitary disease is significantly more frequent in familial MEN1 cases than in the sporadic form of the disease.[12] The prevalence of pituitary adenomas is around 40% in multiple endocrine neoplasia Type I patients, with prolactinomas being the most common type.[22,23] In a large study, Verges et al compared the characteristics of pituitary disease developed in 324 MEN1 patients with those of 110 non-MEN1 patients with pituitary adenomas, matched for age, year of diagnosis and clinical follow-up.[23] Forty-two percent of the MEN1 patients developed pituitary tumors and interestingly, pituitary disease among the familial MEN1 cases was found to be more frequent than in the sporadic MEN1 cases (59% vs 34% respectively). Furthermore, pituitary adenomas were significantly larger and more aggressive in MEN1 cases compared to patients without MEN1. Indeed, 85% of these MEN1-related prolactinomas developed macro adenomas and one fifth was invasive.[23,24] Therapies based on the use of dopamine agonists for these more aggressive MEN1-related prolactinomas showed little or poor response.[23,24] All types of mutation were observed, including frameshifts, nonsenses, missenses, germ-line MEN1 encompassing large deletion, strongly suggesting the absence of any phenotype-genotype correlation.

Together, these studies indicate that the *MEN1* gene plays a critical role in the initiation event of familial pituitary adenomas, particularly prolactinomas and potentially in some sporadic ACTH-producing tumors, nonfunctional adenomas, somatotroph adenomas and prolactinomas. Despite the significant recent progress obtained in understanding the genetic basis of the pathogenesis of pituitary adenomas, these studies also highlight the need for further and better understanding of the role of MEN1 mutations in the initiation of pituitary tumors.

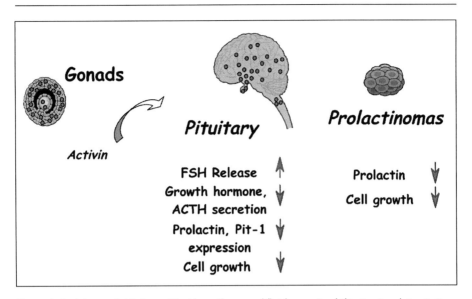

Figure 1. Activin was initially purified from the gonad fluid upon its ability to stimulate pituitary follicle-stimulating hormone (FSH) release from the gonadotropes. Activin can block secretion of ACTH and growth hormone. In addition, activin also inhibits prolactin synthesis and expression of the transcription factor Pit-1 as well as cell growth in normal pituitary lactotrope and pituitary adenomas such as prolactinomas. Activin (green) and its natural antagonist inhibin (purple) are produced in both the gonads and pituitary. A color version of this image is available at www.landesbioscience.com/curie

The Activin/TGF-β Superfamily

The transforming growth factor-β (TGF-β) family is represented by widespread, evolutionary conserved polypeptides that regulate growth, differentiation and apoptosis in nearly all cell types.[25] TGF-β, the prototype of the family and its receptors are expressed by every cell type in the body and defects in its signaling pathway have been implicated in multiple human disorders including cancer.[26] Similarly, activin, that was isolated from gonadal fluid[27,28] and initially recognized for its important role in the regulation of the anterior pituitary function,[29] is a critical regulator of cell growth and hormonal synthesis (Fig. 1).[30] The antiproliferative and more recently discovered pro-apoptotic effects of activin have been observed in many different cell types such as erythroleukemia,[31] capillary endothelial,[32] immune,[33,34] breast cancer,[35-37] hepatocyte[38-43] and pituitary.[5] Truncated activin receptor forms are often found in human pituitary adenomas and function as dominant negative receptors, contributing to pituitary tumorigenesis by blocking the growth inhibitory effect of activin.[44] The activin/TGF-β signal transduction is primarily mediated through the canonical Smad pathway (see Chapter 4 for more details). Upon activation, the Smad protein complex will translocate to the nucleus where it can bind DNA but with a very low affinity.[45] In order to achieve high affinity binding, the Smads associate with various DNA binding partners.[46] It is thought that these partner proteins which act as co-activators or corepressors are functionally expressed in different cell types, thus providing a basis for tissue and cell type specific functions for TGF-β ligands.[47,48] Missense mutations within the carboxyl terminal effector domain of Smad2 and Smad4 have been found in numerous cancers such as pancreatic,[49,50] biliary track,[51] colon,[52-54] lung,[55] head and neck[56] and liver[57] carcinomas, consistent with a role for activin and TGFβ as tumor suppressor.

Table 1. Biological functions of members of the activin/TGF-β family in the anterior pituitary

Biological Functions	Anterior Pituitary Cell Types	Shared Effects from Other Members of the Family
▲FSH[66]	Gonadotropes	▲BMP15[67] ▼Inhibin[68]
▼GH[66]	Somatotropes	
▼ACTH[66]	Corticotropes	
▲Smad7[69]	Gonadotropes Somatotropes Corticotropes Lactotropes	▲TGF-β[69]
▲Follistatin[70]	Gonadotropes	
▼Prolactin[5]	Lactotropes	▼TGF-β[71]
▼Pit-1[60]	Lactotropes	▼TGF-β[60]
▼Cell growth[5]	Lactotropes	▼TGF-β[4]
▲Menin[5]	Lactotropes	▲TGF-β[4]

Activin and other members of the TGF-β family upregulate (▼) or downregulate (▲) expression of various factors in the different cell types of the anterior pituitary gland. Opposite effect between activin and another family member reflects an antagonism between the two growth factors while similar effect reflects synergistic or additive effects between the two members.

Activin/TGF-β in the Pituitary

Pituitary gland function is controlled by a large array of hormones and growth factors. Activin and TGF-β regulate the secretion of a variety of endocrine products[58] and play an important role in regulating anterior pituitary gland function (Table 1). The pituitary action of activin is not restricted to gonadotropes and activin also modulates the function of other pituitary cell types such as the somatotropes and lactotropes. In addition to stimulating FSH release from the gonadotropes, activin inhibits basal growth hormone and adrenocorticotropin secretion.[59] Finally, activin also inhibits expression of the transcription factor Pit-1 in pituitary cells[60] and acts as a negative regulator of prolactin expression and secretion in primary pituitary cells and prolactinomas.[5,61] Consistent with the critical role of activin in cell growth regulation, alterations of the activin signaling pathway, resulting from mutation or truncation of the activin receptors, are associated with human tumors such as pituitary adenomas.[62,63] Indeed, previous work indicated that clinically nonfunctioning pituitary tumors often express alternately spliced activin receptor isoforms[64] that act as suppressors of the activin signaling pathway.[65] Expression screening for such isoforms in tumor samples from patients with prolactinomas did not reveal the presence of abnormal forms for either the Type I or Type II activin receptor. However, it is possible that inactivating alterations of the other downstream components of the activin signaling pathway in human prolactinomas may contribute to tumorigenesis by blocking activin-induced growth arrest and prolactin gene repression.

Activin Inhibits Prolactin Gene Expression and Signalling

The role of activin on prolactin gene expression was also characterized in rat somatolactotrope GH4C1 cells. Prolactin mRNA and protein levels were markedly reduced after activin treatment.[5] Even though we found that part of the activin inhibitory effects on prolactin gene expression were due to down-regulation of the transcription factor Pit-1,[60] we also showed that activin directly regulates prolactin expression at the transcriptional level and exerts a strong inhibition of both the rat and human gene promoter activities.[5] The responsive activin DNA binding elements of the human prolactin gene promoter were further mapped to the region of the promoter proximal to the start site.[5] Interestingly, we recently identified a novel regulatory crosstalk mechanism by which activin/TGF-β-induced Smad signaling acts to antagonize prolactin-mediated signaling and target genes expression.[72] Thus, activin/TGF-β exerts a tight control on prolactin signal transduction by both blocking prolactin expression and antagonizing its downstream signaling cascades.

Menin Inactivation

Twenty-six percent of mice heterozygous for deletion of the *MEN1* gene develop large pituitary tumors by 16 months of age.[18] Interestingly, while the homozygous deletion of *MEN1* is embryonic lethal (at mid-gestation with defects in multiple organs), the *MEN1* heterozygous knockout mice are alive but develop endocrine tumors during their lifetime similar to human MEN1 patients.[18,73,74] Interestingly, inactivation of menin through different antisense technologies (cDNA antisense, oligonucleotide antisense and siRNA) blocks activin and TGF-β signaling in the pituitary and resulted in an increased expression of the hormone prolactin and transcription factor Pit-1 as well as a loss of pituitary cell growth inhibition by activin.[5] Similarly, overexpression of menin, the product of the *MEN1* gene, leads to reduced prolactin expression.[75] Moreover, menin physically interacts with Smad3 in lactotrope cells.[5,76] Thus, menin appears as a novel activin/TGF-β downstream signaling effector molecule and its inactivation leads to the loss of activin and TGF-β responses in the pituitary gland (Fig. 2).

Loss of Menin Inhibits TGF-β Induced Transcriptional Activity

The first evidence of the involvement of menin in the TGFβ signaling pathway came from experiments using various strategies to block menin expression in the rat anterior somatolactotrope GH4C1 cell line.[4] GH4C1 cells are highly differentiated neuroendocrine cells that retain the capacity to synthesize and secrete growth hormone and prolactin in a hormone-regulated manner.[77,78] This cell line was established from rat pituitary tumor cells and is widely used as an in vitro model of pituitary tumors.[5,60,79,80] Interestingly, overexpression of an antisense menin cDNA in these cells antagonized the normal inhibitory effect of TGFß on cell proliferation and cell viability assays.[4] Furthermore, TGF-β transcriptional activity was also blocked as antisense menin expression markedly reduced TGF-β-responsive gene promoter activity.[4] The specificity of the response was demonstrated by the restoration of transcriptional activity with cotransfection of increasing amounts of a sense menin construct.[4]

Menin Interacts with Smad Proteins

Considering the fact that menin is required for TGF-β signaling and that Smad signaling most often requires the recruitment and association of Smad2 and Smad3 with co-activators or corepressors of transcription, it became evident that menin could be a candidate Smad-interacting partner. Indeed, we further demonstrated using co-immunoprecipitation experiments that menin could specifically interact with Smad3 but not Smad2 or Smad4.[4] The interaction was further shown to be direct, using GST-Smad3 pull-down assays with in vitro transcribed/translated full-length and deletion mutants of menin. Further mapping studies will precisely localize the interacting regions in both proteins. Interestingly, this study not only highlights menin a novel Smad-interacting partner but also indicates that Smad2 and Smad3 signaling may differ in the anterior pituitary. Menin was later shown to interact with the bone morphogenetic protein (BMP)-regulated Smads.[81] Thus, menin appears a central regulator of Smad signaling relaying the signal transduction pathways of various TGF-β family members (Fig. 2).

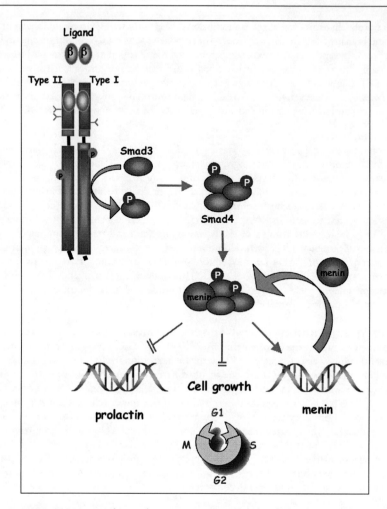

Figure 2. Activin/TGF-β signal transduction: Activin/TGF-β binding to a serine kinase Type II receptor is followed by recruitment and transphosphorylation of the Type I receptor, thereby activating its kinase activity which then recruits and phosphorylates the Smads. In pituitary adenoma cells the Smad complex recruits the tumor suppressor menin to inhibit cell growth, in the G1 phase of the cell cycle and to repress prolactin expression. Expression of menin itself is upregulated by activin/TGF-β and Smad signaling, thus menin acts in a positive feedback loop manner.

Smads and Menin Are Required for Activin-Mediated Cell Growth Inhibition and Repression of Prolactin Gene Expression

The requirement of the canonical Smad pathway was demonstrated using overexpression of the inhibitory Smad7 which resulted in a complete reversal of activin-induced prolactin inhibition. Furthermore, overexpression of two dominant-negative forms of Smad2 (ΔNSmad2) and Smad3 (ΔNSmad3) in which the two C-terminal serine residues, the target of Smad phosphorylation by the Type I receptor, are mutated to alanine[82,83] also led to reversal of the activin effects.[5] The Smads usually act by recruiting co-activators or corepressors to mediate their transcriptional effects. Pleiades of Smad-interacting molecules have been identified, often in cell and tissue specific manner. As mentioned above, menin is a Smad3-interacting partner that is required to mediate

TGF-β signaling in pituitary cells.[4] Moreover, overexpression of menin in pituitary cells inhibits prolactin gene expression[75] and, as illustrated in Figure 2, our group recently found that menin is also required for activin to repress prolactin expression.[5] Indeed, reducing menin expression levels using different siRNAs corresponding to the rat *MEN1* gene is sufficient to block activin-mediated prolactin inhibition, demonstrating the critical role played by menin in activin signal transduction in anterior pituitary cells.[5] Activin, like TGF-β, is known to induce growth inhibition of epithelial, endothelial, lymphoid and hematopoietic cells. Both activin and TGF-β efficiently reduce cell growth of GH4C1 lactotrope cells and that these effects were also strictly dependent on menin, as blocking menin expression resulted in the loss of growth inhibitory signals by both activin and TGF-β.[4,5] These data demonstrate that activin acts as a potent pituitary tumor cell growth inhibitor and that this effect requires the tumor suppressor menin.

Conclusion

Menin plays an important role in supporting activin, TGF-β and Smad3 transcriptional control of cell growth and hormonal synthesis and reduced menin expression disrupts activin/TGF-β-mediated prolactin expression and pituitary tumor cell growth inhibition. As mentioned earlier, Smad proteins often recruit co-activators but also corepressors of transcription such as the oncoproteins Ski and SnoN, to mediate their effects. It is therefore conceivable that menin prevents Smad association with such Smad transcriptional corepressors.[84,85] In such a model menin would act by blocking the effect of these corepressors, thereby facilitating activin and TGF-β-mediated inhibition of cell growth and prolactin expression. Pituitary adenomas are common, but in contrast to sporadic parathyroid and enteropancreatic tumors, mutation of the *MEN1* gene is not a major contributor to sporadic pituitary tumorigenesis. However, variable and decreased expression of the menin protein was noted in a series of sporadic pituitary adenomas.[21] Therefore, it is likely that reduced menin expression is contributing to the more common sporadic pituitary tumorigenesis. Interestingly, it was also recently discovered that both activin and TGF-β could stimulate menin expression in a rapid and dose-dependent manner,[4,5] thus suggesting that menin could act in a positive feedback loop manner downstream of these growth factor receptors (Fig. 2). The transcriptional machinery by which activin/TGF-β regulates menin expression remains to be elucidated and future studies using the menin gene promoter will undoubtedly be insightful.

In summary, these studies shed light on the mechanisms by which activin and TGF-β regulate hormone expression and cell growth arrest in the anterior pituitary, through the Smad pathway and the tumor suppressor menin. Further detailed understanding of how menin blocks Smad signaling, identification of the transcriptional machinery by which activin/TGF-β up-regulate menin protein levels and development of new drugs that will mimic menin and Smad signaling in the pituitary will be helpful to offer new treatment avenues for patients with pituitary adenomas.

Acknowledgements

The author wishes to acknowledge the support of the Canadian Institutes for Health Research (MOP-53141). Jean-Jacques Lebrun currently holds a Research Scientist Award from the National Cancer Institute of Canada supported with funds provided by the Canadian Cancer Society.

References

1. Heaney AP, Melmed S. Molecular targets in pituitary tumours. Nat Rev Cancer 2004; 4:285-295.
2. Ben-Jonathan N, Hnasko R. Dopamine as a prolactin (PRL) inhibitor. Endocr Rev 2001; 22:724-763.
3. Asa SL, Ezzat S. The cytogenesis and pathogenesis of pituitary adenomas. Endocr Rev 1998; 19:798-827.
4. Kaji H, Canaff L, Lebrun JJ et al. Inactivation of menin, a Smad3-interacting protein, blocks transforming growth factor type β signaling. Proc Natl Acad Sci USA 2001; 98:3837-3842.
5. Lacerte A, Lee EH, Reynaud R et al. Activin inhibits pituitary prolactin expression and cell growth through Smads, Pit-1 and menin. Mol Endocrinol 2004; 18:1558-1569.
6. Faglia G. Epidemiology and pathogenesis of pituitary adenomas. Acta Endocrinol (Copenh) 1993; 129 (Suppl 1):1-5.

7. Shimon I, Melmed S. Genetic basis of endocrine disease: pituitary tumor pathogenesis. J Clin Endocrinol Metab 1997; 82:1675-1681.
8. Spada A, Vallar L, Faglia G. G-proteins and hormonal signalling in human pituitary tumors: genetic mutations and functional alterations. Frontiers in Neuroendocrinology 1993; 14:214-232.
9. Ledent C, Parma J, Pirson I et al. Positive control of proliferation by the cyclic AMP cascade: an oncogenic mechanism of hyper-functional adenoma. J Endocrinol Invest 1995; 18:120-122.
10. Weil RJ, Vortmeyer AO, Huang S et al. 11q13 allelic loss in pituitary tumors in patients with multiple endocrine neoplasia syndrome type 1. Clin Cancer Res 1998; 4:1673-1678.
11. Eubanks PJ, Sawicki MP, Samara GJ et al. Putative tumor-suppressor gene on chromosome 11 is important in sporadic endocrine tumor formation. Am J Surg 1994; 167:180-185.
12. Wautot V, Vercherat C, Lespinasse J et al. Germline mutation profile of MEN1 in multiple endocrine neoplasia type 1: search for correlation between phenotype and the functional domains of the MEN1 protein. Hum Mutat 2002; 20:35-47.
13. Boggild MD, Jenkinson S, Pistorello M et al. Molecular genetic studies of sporadic pituitary tumors. J Clin Endocrinol Metab 1994; 78:387-392.
14. Thakker RV, Pook MA, Wooding C et al. Association of somatotrophinomas with loss of alleles on chromosome 11 and with gsp mutations. J Clin Invest 1993; 91:2815-2821.
15. Herman V, Fagin J, Gonsky R et al. Clonal origin of pituitary adenomas. J Clin Endocrinol Metab 1990; 71:1427-1433.
16. Bates AS, Farrell WE, Bicknell EJ et al. Allelic deletion in pituitary adenomas reflects aggressive biological activity and has potential value as a prognostic marker. J Clin Endocrinol Metab 1997; 82:818-824.
17. Dong Q, Debelenko LV, Chandrasekharappa SC et al. Loss of heterozygosity at 11q13: analysis of pituitary tumors, lung carcinoids, lipomas and other uncommon tumors in subjects with familial multiple endocrine neoplasia type 1. J Clin Endocrinol Metab 1997; 82:1416-1420.
18. Crabtree JS, Scacheri PC, Ward JM et al. A mouse model of multiple endocrine neoplasia, type 1, develops multiple endocrine tumors. Proc Natl Acad Sci USA 2001; 98:1118-1123.
19. Zhuang Z, Ezzat SZ, Vortmeyer AO et al. Mutations of the MEN1 tumor suppressor gene in pituitary tumors. Cancer Res 1997; 57:5446-5451.
20. Poncin J, Stevenaert A, Beckers A. Somatic MEN1 gene mutation does not contribute significantly to sporadic pituitary tumorigenesis. Eur J Endocrinol 1999; 140:573-576.
21. Theodoropoulou M, Cavallari I, Barzon L et al. Differential expression of menin in sporadic pituitary adenomas. Endocr Relat Cancer 2004; 11:333-344.
22. Burgess J, Shepherd J, Parameswaran V et al. Prolactinomas in a large kindred with multiple endocrine neoplasia type 1: clinical features and inheritance pattern. J Clin Endocrinol Metab 1996; 81:1841-1845.
23. Verges B, Boureille F, Goudet P et al. Pituitary disease in MEN type 1 (MEN1): data from the France-Belgium MEN1 multicenter study. J Clin Endocrinol Metab 2002; 87:457-465.
24. Beckers A, Betea D, Socin HV et al. The treatment of sporadic versus MEN1-related pituitary adenomas. J Intern Med 2003; 253:599-605.
25. Massague J. The transforming growth factor-β family. Annu Rev Cell Biol 1990; 6:597-641.
26. Massague J. TGF-β signal transduction. Annu Rev Biochem 1998; 67:753-791.
27. Vale W, Rivier J, Vaughan J et al. Purification and characterization of an FSH releasing protein from porcine ovarian follicular fluid. Nature 1986; 321:776-779.
28. Ling N, Ying SY, Ueno N et al. Pituitary FSH is released by a heterodimer of the β-subunits from the two forms of inhibin. Nature 1986; 321:779-782.
29. Vale W, Rivier C, Hsueh A et al. Chemical and biological characterization of the inhibin family of protein hormones. Recent Prog Horm Res 1988; 44:1-34.
30. Chen YG, Lui HM, Lin SL et al. Regulation of cell proliferation, apoptosis and carcinogenesis by activin. Exp Biol Med (Maywood) 2002; 227:75-87.
31. Lebrun JJ, Vale WW. Activin and inhibin have antagonistic effects on ligand-dependent heteromerization of the type I and type II activin receptors and human erythroid differentiation. Mol Cell Biol 1997; 17:1682-1691.
32. McCarthy SA, Bicknell R. Inhibition of vascular endothelial cell growth by activin-A. J Biol Chem 1993; 268:23066-23071.
33. Brosh N, Sternberg D, Honigwachs-Sha'anani J et al. The plasmacytoma growth inhibitor restrictin-P is an antagonist of interleukin 6 and interleukin 11. Identification as a stroma-derived activin A. J Biol Chem 1995; 270:29594-29600.
34. Valderrama-Carvajal H, Cocolakis E, Lacerte A et al. Activin/TGF-β induce apoptosis through Smad-dependent expression of the lipid phosphatase SHIP. Nat Cell Biol 2002; 4:963-969.

35. Kalkhoven E, Roelen BA, de Winter JP et al. Resistance to transforming growth factor β and activin due to reduced receptor expression in human breast tumor cell lines. Cell Growth Differ 1995; 6:1151-1161.
36. Liu QY, Niranjan B, Gomes P et al. Inhibitory effects of activin on the growth and morpholgenesis of primary and transformed mammary epithelial cells. Cancer Res 1996; 56:1155-1163.
37. Cocolakis E, Lemay S, Ali S et al. The p38 MAPK pathway is required for cell growth inhibition of human breast cancer cells in response to activin. J Biol Chem 2001; 276:18430-18436.
38. Yasuda H, Mine T, Shibata H et al. Activin A: an autocrine inhibitor of initiation of DNA synthesis in rat hepatocytes. J Clin Invest 1993; 92:1491-1496.
39. Xu J, McKeehan K, Matsuzaki K et al. Inhibin antagonizes inhibition of liver cell growth by activin by a dominant-negative mechanism. J Biol Chem 1995; 270:6308-6313.
40. Zauberman A, Oren M, Zipori D. Involvement of p21(WAF1/Cip1), CDK4 and Rb in activin A mediated signaling leading to hepatoma cell growth inhibition. Oncogene 1997; 15:1705-1711.
41. Takabe K, Lebrun JJ, Nagashima Y et al. Interruption of activin A autocrine regulation by antisense oligodeoxynucleotides accelerates liver tumor cell proliferation. Endocrinology 1999; 140:3125-3132.
42. Chen W, Woodruff TK, Mayo KE. Activin A-induced HepG2 liver cell apoptosis: involvement of activin receptors and smad proteins. Endocrinology 2000; 141:1263-1272.
43. Ho J, de Guise C, Kim C et al. Activin induces hepatocyte cell growth arrest through induction of the cyclin-dependent kinase inhibitor p15INK4B and Sp1. Cell Signal 2004; 16:693-701.
44. Zhou Y, Sun H, Danila DC et al. Truncated activin type I receptor Alk4 isoforms are dominant negative receptors inhibiting activin signaling. Mol Endocrinol 2000; 14:2066-2075.
45. Shi Y, Wang YF, Jayaraman L et al. Crystal structure of a Smad MH1 domain bound to DNA: insights on DNA binding in TGF-β signaling. Cell 1998; 94:585-594.
46. Massague J, Wotton D. Transcriptional control by the TGF-β/Smad signaling system. EMBO J 2000; 19:1745-1754.
47. Chen X, Weisberg E, Fridmacher V et al. Smad4 and FAST-1 in the assembly of activin-responsive factor. Nature 1997; 389:85-89.
48. Hata A, Seoane J, Lagna G et al. OAZ uses distinct DNA- and protein-binding zinc fingers in separate BMP- Smad and Olf signaling pathways. Cell 2000; 100:229-240.
49. Hahn SA, Hoque AT, Moskaluk CA et al. Homozygous deletion map at 18q21.1 in pancreatic cancer. Cancer Res 1996; 56: 490-494.
50. Moskaluk CA, Kern SE. Cancer gets Mad: DPC4 and other TGFβ pathway genes in human cancer. Biochim Biophys Acta 1996; 1288:M31-33.
51. Hahn SA, Bartsch D, Schroers A et al. Mutations of the DPC4/Smad4 gene in biliary tract carcinoma. Cancer Res 1998; 58:1124-1126.
52. Eppert K, Scherer SW, Ozcelik H et al. MADR2 maps to 18q21 and encodes a TGFβ-regulated MAD-related protein that is functionally mutated in colorectal carcinoma. Cell 1996; 86:543-552.
53. Riggins GJ, Kinzler KW, Vogelstein B et al. Frequency of Smad gene mutations in human cancers. Cancer Res 1997; 57:2578-2580.
54. Takagi Y, Koumura H, Futamura M et al. Somatic alterations of the SMAD-2 gene in human colorectal cancers. Br J Cancer 1998; 78:1152-1155.
55. Uchida K, Nagatake M, Osada H et al. Somatic in vivo alterations of the JV18-1 gene at 18q21 in human lung cancers. Cancer Res 1996; 56:5583-5585.
56. Kim SK, Fan Y, Papadimitrakopoulou V et al. DPC4, a candidate tumor suppressor gene, is altered infrequently in head and neck squamous cell carcinoma. Cancer Res 1996; 56:2519-2521.
57. Yakicier MC, Irmak MB, Romano A et al. Smad2 and Smad4 gene mutations in hepatocellular carcinoma. Oncogene 1999; 18:4879-4883.
58. Luisi S, Florio P, Reis F et al. Expression and secretion of activin A: possible physiological and clinical implications. Eur J Endocrinology 2001; 145:225-236.
59. Lebrun JJ, Chen Y, Vale WW. Receptor serine kinases and signaling by activins and inhibins. In: Aono T, Sugino H Vale WW eds. Inhibin, activin and follistatin: Regulatory functions in system and cell biology. New York: Springer-Verlag, 1997:1-21.
60. De Guise C, Lacerte A, Rafiei S et al. Activin inhibits the human Pit-1 gene promoter through the p38 kinase pathway in a Smad-independent manner. Endocrinology 2006; 147:4351-4362.
61. Murata T, Ying S. Transforming growth factor-β and activin inhibit basal secretion of prolactin in a pituitary monolayer culture system. Proc Soc Exp Biol Med 1991; 198:599-605.
62. Danila DC, Inder WJ, Zhang X et al. Activin effects on neoplastic proliferation of human pituitary tumors. J Clin Endocrinol Metab 2000; 85:1009-1015.
63. Su GH, Bansal R, Murphy KM et al. ACVR1B (ALK4, activin receptor type 1B) gene mutations in pancreatic carcinoma. Proc Natl Acad Sci USA 2001; 98:3254-3257.

64. Alexander JM, Bikkal HA, Zervas NT et al. Tumor-specific expression and alternate splicing of messenger ribonucleic acid encoding activin/transforming growth factor-β receptors in human pituitary adenomas. J Clin Endocrinol Metab 1996; 81:783-790.
65. Zhou Y, Sun H, Danila DC et al. Truncated activin type I receptor Alk4 isoforms are dominant negative receptors inhibiting activin signaling. Mol Endocrinol 2000; 14:2066-2075.
66. Bilezikjian LM, Blount AL, Corrigan AZ et al. Actions of activins, inhibins and follistatins: implications in anterior pituitary function. Clin Exp Pharmacol Physiol 2001; 28:244-248.
67. Otsuka F, Shimasaki S. A novel function of bone morphogenetic protein-15 in the pituitary: selective synthesis and secretion of FSH by gonadotropes. Endocrinology 2002; 143:4938-4941.
68. Lewis KA, Gray PC, Blount AL et al. βglycan binds inhibin and can mediate functional antagonism of activin signalling. Nature 2000; 404:411-414.
69. Bilezikjian LM, Corrigan AZ, Blount AL et al. Regulation and actions of Smad7 in the modulation of activin, inhibin and transforming growth factor-β signaling in anterior pituitary cells. Endocrinology 2001; 142:1065-1072.
70. Bilezikjian LM, Corrigan AZ, Blount AL et al. Pituitary follistatin and inhibin subunit messenger ribonucleic acid levels are differentially regulated by local and hormonal factors. Endocrinology 1996; 137:4277-4284.
71. Abraham EJ, Faught WJ, Frawley LS. Transforming growth factor β1 is a paracrine inhibitor of prolactin gene expression. Endocrinology 1998; 139:5174-5181.
72. Cocolakis E, Dai M, Drevet L et al. Smad signaling antagonizes STAT5-mediated gene transcription and mammary epithelial cell differentiation. J Biol Chem 2008; 283:1293-1307.
73. Bertolino P, Tong WM, Galendo D et al. Heterozygous Men1 mutant mice develop a range of endocrine tumors mimicking multiple endocrine neoplasia type 1. Mol Endocrinol 2003; 17:1880-1892.
74. Bertolino P, Radovanovic I, Casse H et al. Genetic ablation of the tumor suppressor menin causes lethality at mid-gestation with defects in multiple organs. Mech Dev 2003; 120:549-560.
75. Namihira H, Sato M, Murao K et al. The multiple endocrine neoplasia type 1 gene product, menin, inhibits the human prolactin promoter activity. Journal of Molecular Endocrinology 2002; 29:297-304.
76. Kaji H, Canaff L, Lebrun J-J et al. Inactivation of menin, a Smad3-interacting protein, blocks transforming growth factor type β signaling. Proc Natl Acad Sci USA 2001; 98:3837-3842.
77. Keech C, Gutierrez-Hartmann A. Analysis of rat prolactin promoter sequences that mediate pituitary-specific and 3', 5'-cyclic adenosine monophosphate-regulated gene expression in vivo. Mol Endocrinol 1989; 3:832-839.
78. Keech C, Gutierrez-Hartmann A. Insulin activation of rat prolactin promoter activity. Mol Cell Endocrinol 1991; 78:55-60.
79. Tashjian AH Jr, Yasumura Y, Levine L et al. Establishment of clonal strains of rat pituitary tumor cells that secrete growth hormone. Endocrinology 1968; 82:342-352.
80. Song JY, Jin L, Lloyd RV. Effects of estradiol on prolactin and growth hormone messenger RNAs in cultured normal and neoplastic (MtT/W15 and GH3) rat pituitary cells. Cancer Res 1989; 49:1247-1253.
81. Sowa H, Kaji H, Hendy GN et al. Menin is required for bone morphogenetic protein 2- and transforming growth factor β-regulated osteoblastic differentiation through interaction with Smads and Runx2. J Biol Chem 2004; 279:40267-40275.
82. Macias-Silva M, Abdollah S, Hoodless PA et al. MADR2 is a substrate of the TGFβ receptor and its phosphorylation is required for nuclear accumulation and signaling. Cell 1996; 87:1215-1224.
83. Liu X, Sun Y, Constantinescu SN et al. Transforming growth factor β-induced phosphorylation of Smad3 is required for growth inhibition and transcriptional induction in epithelial cells. Proc Natl Acad Sci USA 1997; 94:10669-10674.
84. Stroschein SL, Wang W, Zhou S et al. Negative feedback regulation of TGF-β signaling by the SnoN oncoprotein. Science 1999; 286:771-774.
85. Luo K, Stroschein SL, Wang W et al. The Ski oncoprotein interacts with the Smad proteins to repress TGFβ signaling. Genes Dev 1999; 13:2196-2206.

Chapter 8

The Role of Menin in Parathyroid Tumorigenesis

Colin Davenport and Amar Agha*

Abstract

Primary hyperparathyroidism is a common disorder that involves the pathological enlargement of one or more parathyroid glands resulting in excessive production of parathyroid hormone (PTH). The exact pathogenesis of this disease remains to be fully understood. In recent years interest has focussed on the interaction between menin protein and the transforming growth factor (TGF)-β/Smad signalling pathway. In vitro experimentation has demonstrated that the presence of menin is required for TGF-β to effectively inhibit parathyroid cell proliferation and PTH production. This observation correlates with the almost universal occurrence of parathyroid tumors accompanying the inactivation of menin in multiple endocrine neoplasia Type 1 (MEN 1) syndrome and the high rate of somatic menin gene mutations seen in sporadic parathyroid adenomas. This chapter aims to review the role of menin in primary hyperparathyroidism and parathyroid hormone-regulation, including the influences of *MEN1* gene mutations on parathyroid cell proliferation, differentiation and tumorigenesis.

Introduction

Primary hyperparathyroidism is the commonest cause of hypercalcemia in the outpatient population, with an incidence of approximately 25 per 100,000.[1] It is typically a disease of the middle aged to elderly and women are affected four times more commonly than men. The majority of cases are caused by sporadic enlargement of a single parathyroid gland (80-85%). Multi-glandular involvement occurs in such familial syndromes as multiple endocrine neoplasia (MEN) syndrome (Type I or IIa), familial isolated hyperparathyroidism (FIHP), hyperparathyroidism jaw tumor syndrome and familial hypocalciuric hypercalcemia (Table 1).[2]

To date, a number of genes and their protein products have been implicated in the process of parathyroid tumorigenesis. These include activation of the proto-oncogene cyclin D1,[3] somatic mutations of mitochondrial deoxyribonucleic acid (mtDNA)[4] and the loss of the chromosome 13q fragment.[5] Despite ongoing investigation into these and other proposed mechanisms, the exact molecular pathogenesis remains unclear. However, in recent years much research has focused on the role of the menin protein in the development of parathyroid tumors, which has significantly added to our understanding of their origin and will be the subject of discussion in this chapter.

The *MEN1* Gene

The *MEN1* gene is located on chromosome 11q13[6] and consists of 10 exons that encode a protein named menin. This 610-amino acid protein resides primarily in the nucleus and interacts

*Corresponding Author: Amar Agha—Academic Department of Endocrinology, Beaumont Hospital and Royal College of Surgeons in Ireland Medical School, Beaumont Road, Dublin 9, Ireland. Email: amaragha@beaumont.ie

SuperMEN1: Pituitary, Parathyroid and Pancreas, edited by Katalin Balogh and Attila Patocs. ©2009 Landes Bioscience and Springer Science+Business Media.

Table 1. Comparison of sporadic and familial causes of hyperparathyroidism

Disease	Inheritance	Age at Onset of Hypercalcaemia (Yrs)	No of Abnormal Parathyroid Glands	Pathophysiology	Histology
Sporadic adenoma	Not inherited	50-60	1, very enlarged	Stepwise carcinogenesis involving multiple genes such as MEN1 results in a neoplasm	Encapsulated, cellular, homogenous lesion, mainly composed of chief cells, loss of fat, minimal mitotic activity, no invasion of surrounding tissue
MEN1 syndrome	Autosomal dominant	20-25	Multiple, moderately enlarged	Germline mutation in MEN1 combines with an acquired mutation of the other allele	As above, multiple glands
MEN2A syndrome	Autosomal dominant	25-35	May be multiple, moderately enlarged	Mutations of the ret proto-oncogene	As above
Hyperparathyroidism jaw tumor syndrome	Autosomal dominant	30-40	Multiple, moderately enlarged	Inactivating mutations of the HRPT2 tumor suppressor gene	Monoclonal expansion, cysts, malignant change
Sporadic parathyroid cancer	Not inherited	40-50	1, moderately enlarged	Associated with loss HRPT2 activity	Circumscribed; gray-white, firm, irregular mass, dense fibrous bands, spindled tumor cells, vascular and perineural invasion
Familial hypocalciuric hypercalcaemia	Autosomal dominant	Birth	Multiple, minimally enlarged	Inactivation of the calcium sensing receptor genes in the parathyroid and the renal tubules	Histologically normal

with several transcription regulators such as JunD, NF-κB and the Smad family of proteins. In doing so, menin exerts a potent tumor suppressor effect on several endocrine tissues. Individuals who have a germline inactivating mutation of this gene develop MEN1 syndrome. In accordance with Knudson's two-hit hypothesis, the germline *MEN1* mutations combine with acquired somatic mutations of the second copy of *MEN1* (Fig. 1). This leads to monoclonal expansion and multiple neoplasia arising in such organs as the pituitary, parathyroids and the endocrine pancreas among others in an autosomal dominant manner. Of all these endocrinopathies, hyperparathyroidism is the commonest manifestation of the syndrome, with a typical age of onset of 20-25 years and an almost 100% penetrance by age 50.[7] *MEN1* patients generally develop adenomas arising from three or even all four glands.

In addition to MEN1 syndrome, when sporadic parathyroid adenomas are subjected to genetic analysis, a significant percentage will demonstrate acquired biallelic inactivating mutations of the *MEN1* gene, as has been demonstrated by Heppner et al[8-9] Still more of these sporadic cases demonstrate other mutations in the 11q region which may have detrimental effects on the function of the *MEN1* gene that have yet to be elucidated.[10] Thus it has been suggested that loss

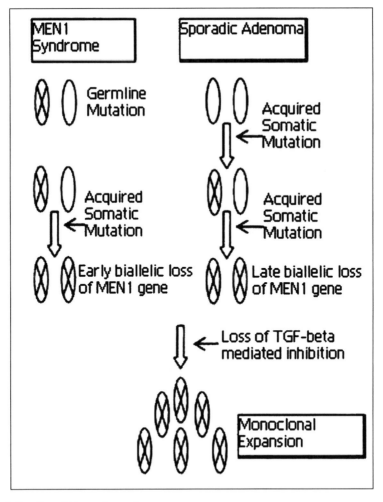

Figure 1. Parathyroid tumorigenesis.

of menin may play a vital role in a significant percentage of parathyroid tumors. In this chapter, we will focus on the genetics of *MEN1* mutations, the mechanism by which menin exerts its tumor suppressor effect and the development of parathyroid tumors and hyperparathyroidism as a consequence of its inactivation.

MEN1 Related Mutations and Menin Expression in Hereditary and Sporadic Hyperparathyroidism (Genotype-Phenotype Correlation)

Since the sequencing of the *MEN1* gene in 1997, there have been 1,133 germline mutations identified. These consist of 23% nonsense mutations, 9% splice site mutations, 41% frameshift deletions or insertions, 6% in-frame deletions or insertions, 20% missense mutations and 1% whole or partial gene deletions. In a recent review of all identified mutations by Lemos et al, several mutations were found to recur in apparently unrelated kindreds, thereby indicating potential mutational hot spots for the syndrome.[11] The authors noted that mutations at four sites accounted for 12.3% of all mutations (c.249_252delGTCT, deletion at codons 83-84; c.1546_1547insC, insertion at codon 516; c.1378C4T (Arg460Ter); and c.628_631delACAG, deletion at codons 210-211).

However, establishing precise genotype-phenotype correlations in any of these patient populations has, to date, proven difficult. For example, FIHP has a phenotype distinct from MEN1 syndrome and also displays a significantly increased percentage of missense mutations (38% versus 20%, $p < 0.01$) when compared to the latter condition. Furthermore, the Burin variant of MEN1 (frequent occurrence of prolactinomas), first noted in four kindreds in Newfoundland, has been associated with specific nonsense mutations (Tyr312Ter and Arg460Ter) and a Tasmanian variant characterized by an absence of somatotrophinomas has been linked to a specific splice site mutation (c.446-3C4G). However, the unequivocal presence of protein truncating mutations and deletions in some FIHP patients, which are identical to those observed in MEN1 patients, argue against a simple genotype-phenotype correlation and suggest a more complex relationship. Thus, while some subtypes of MEN1 do display distinct phenotypes, the use of genotyping to predict phenotypic expression in the majority of individuals with MEN1 is not yet possible.

With regards to sporadic hyperparathyroidism, as shown by Heppner et al, at least 20% of sporadic parathyroid tumors will demonstrate biallelic inactivating somatic mutations of the *MEN1* gene,[8] and over 200 different somatic mutations of *MEN1* have been identified in these tumors. 18% of these are nonsense mutations, 40% are frameshift deletions or insertions, 6% are in-frame deletions or insertions, 7% are splice-site mutations and 29% are missense mutations. Interestingly, when comparing the common locations of somatic versus germline mutations, a higher frequency of somatic mutations in exon 2 (39 versus 23%, $p < 0.001$) has been found, but the significance of this difference remains unclear.

While 20% of sporadic parathyroid adenomas have demonstrable inactivating mutations, up to 38% show loss of heterozygosity (LOH) of 11q13, with the additional 18% possibly carrying mutations that are not screened for, such as promoter or intron alterations (although as LOH at 11q13 only occurs in 38% of cases, other genes apart from *MEN1* are likely to play a significant role in tumorigenesis in this group). In addition to this population, it is notable that between 10-30% of patients with an MEN1 phenotype will fail to show demonstrable germline mutations in the coding regions or adjacent splice sites of *MEN1*. These patients may also possess promoter or intron mutations that are not identifiable by the present analytical methods.[12] One potential method for identifying a decrease in *MEN1* function in this patient population is by assessing menin expression in parathyroid tissue and to date a limited amount of data has emerged regarding this approach.

In 2000, Bhuiyan et al used RT-PCR and western blotting techniques to compare menin expression in parathyroid tumors from MEN1, primary hyperparathyroidism and secondary hyperparathyroidism patient populations.[13] The authors found a similar level of menin expression in tumors from patients with primary and secondary hyperparathyroidism. As reviewed later in this chapter, one would not expect a reduction in menin expression in secondary hyperparathyroidism, but a significant percentage of primary hyperparathyroid patients would be expected to demonstrate loss of menin expression. The findings in this study may be explained by the fact that no somatic mutations were identified in the primary hyperparathyroidism cohort used. The authors also found a similar level of menin expression in primary hyperparathyroid patients compared to MEN1 patients with missense mutations and a significantly low expression of MEN1 in patients with nonsense or deletion mutations. Applying these findings to an MEN1 phenotype patient with no identifiable germline mutation, Naito et al demonstrated a significantly reduced expression of menin in the parathyroid tissue and a reduced responsiveness to the inhibitory effects of TGF-β.[12] Thus, depending on the MEN1 mutation, differences in menin expression may serve to identify true MEN1 patients in which a germline mutation cannot be found.

TGF-β/Smad3 Signalling

When TGF-β arrives at the cellular membrane, it binds to and activates the Type II serine/threonine kinase receptor. Once activated, this receptor-ligand complex leads to phosphorylation of the Type I receptor, which phosphorylates in turn Smad2 and Smad3. Phosphorylated Smad2 or Smad3 then associates with Smad4 and the resultant complex translocates into the nucleus. Once in the nucleus, the Smad complex binds to promoter elements to up- or down regulate transcriptional activity.[14] As the dominant influence of the TGF-β pathway is the inhibition of cell growth, such as in parathyroid cells, disruption of any aspect of the above pathway may permit inappropriate cell growth and consequent neoplasia development. A comprehensive review of this pathway is available in Chapter 4.

Menin and TGF-β Signalling

The role of menin in the TGF-β pathway was first explored by Kaji et al in 2001.[15] Using a human hepatoma cell line and antisense menin transfection, it was demonstrated that reduction in menin expression on immunoblotting was associated with a significant reduction in TGF-β induced transcriptional activity. The specific point of interaction between menin and the TGF-β pathway was then investigated further using co-immunoprecipitation analyses on monkey-kidney and hamster-ovary cells. A functional interaction between menin and Smad3, but not Smad 2 or 4, was demonstrated and the absence of menin did not reduce the formation of the Smad3/Smad4 complex or its translocation to the nucleus. Taken together, this information indicated that the area of interest lay in the actual binding of Smad3 (as part of the Smad3/Smad4 complex) to DNA. Accordingly, levels of the Smad3/DNA complex were significantly reduced in rat anterior pituitary cells when menin production was blocked. Thus it is felt that the tumor suppressive effect of menin occurs via facilitation of Smad3/DNA binding, which is necessary for successful activation of the TGF-β pathway. (Table 2)

With the above findings derived from a variety of different cell lines, both human and animal, replication of the data using human parathyroid tissue was necessary to extend the hypothesis to primary hyperparathyroidism. To this end, Sowa et al took parathyroid tissue from uremic patients with secondary hyperparathyroidism in order to assess its proliferative and endocrine responses to TGF-β administration.[10] Menin inactivation was accomplished with the use of antisense menin oligonucleotides. The first finding of note was that administration of TGF-β diminished markers of cellular proliferation, including thymidine incorporation into the parathyroid cells, 3-(4,5-Dimethylthiazol-2-yl)-2,5-diphenyltetrazolium Bromide (MTT) dye assay and immunoblot and immunocytochemical analyses of proliferating cell nuclear antigen (PCNA) expression. This observed effect confirmed the role of TGF-β as an inhibitor of cellular growth in parathyroid endocrine tissue. When menin was inactivated, this inhibitory effect was lost and cellular proliferation, as measured by the above markers, proceeded unimpeded. This study also assessed tissue parathyroid hormone (PTH) expression and PTH levels in the cell culture and with menin intact, TGF-β produced a significant reduction in both measures of PTH production. With menin inactivated, an actual increase in the basal levels of PTH expression and secretion was noted despite TGF-β administration. Finally, tissue from a parathyroid adenoma in a patient

Table 2. Evidence for the role of menin in the function of the TGF-β pathway

Paper	Cell Line	Results
Kaji et al (2001)	Human hepatoma cells and antisense menin transfection	Reduction in menin expression on immunoblotting is associated with a reduction in TGF-β induced transcriptional activity
	Monkey-kidney and hamster-ovary cells	A functional interaction exists between menin and Smad3, but not other Smad proteins
		The absence of menin did not reduce the formation of the Smad3/Smad4 complex or its translocation to the nucleus
	Rat anterior pituitary cells	The absence of menin resulted in significant reductions in the formation of Smad3/DNA complex
Sowa et al (2004)	Uremic human parathyroid cells	Administration of TGF-β diminished PTH production and markers of cellular proliferation
		When menin was inactivated, administration of TGF-β produced a small increase in PTH production and cellular proliferation proceeded unimpeded
	MEN1 human parathyroid adenoma cells	Administration of TGF-β had no effect on PTH production or cellular proliferation

with MEN1 syndrome was sampled and TGF-β was added. In this case, with menin effectively absent at the onset of the experiment, the inhibitory cytokine had no effect on cellular proliferation or PTH production.

Following on from the above findings, the Smad3 gene and protein have become the focus of additional research into parathyroid tumorigenesis. The Smad3 gene has been localised to chromosome 15q and a significant number of parathyroid tumors have shown loss of heterozygosity in the region close to this locus.[16] However no specific acquired mutation has been identified in these tumors to date. Interestingly, a link between Smad3 and vitamin D has been demonstrated, with Smad3 also acting as a transcriptional co-activator of the latter hormone's receptor.[17-18] Thus it may be hypothesised that loss of Smad3 function may be associated with loss of vitamin D mediated inhibition of parathyroid function.

Other Forms of Parathyroid Tumorigenesis

Physiologically appropriate hyperparathyroidism, such as that occurring secondary to hypocalcemia, can gain autonomy and become independent of physiological stimuli. If this event occurs, the resultant disorder is termed refractory secondary hyperparathyroidism. Initially it was felt that this progression represented a polyclonal response to generalized growth stimuli, but in studies looking at X-chromosome inactivation in uremic patients with hyperparathyroidism, a 64% incidence of monoclonality was demonstrated.[19] The impaired capacity for DNA repair demonstrated in patients with chronic renal failure has been ventured as one potential reason for the emergence of these monoclonal cell lines.[20] Menin, however, appears to play a minor role, if any, in this cohort. When Shan et al analysed 20 patients with uremia and refractory parathyroid hyperplasia, they found a 75% incidence of monoclonality, but no mutations in the *MEN1* gene locus.[21] Ultimately, although some cases of *MEN1* deletion have been described in secondary hyperplasia, loss of functioning menin does not appear to play an important role in the early stages of this disorder.[22-23]

With regards to parathyroid carcinoma, traditionally this tumor has not been considered as part of the MEN1 syndrome. Instead, a high incidence of inactivating mutations of the *HRPT2* gene, a component of hyperparathyroidism jaw tumor syndrome, has been identified.[24] However, in recent years a small number of case reports have described patients presenting with parathyroid malignancies in conjunction with the clinical features of MEN1 syndrome.[25-26] Additionally, in 2007 Haven et al analysed 23 cases of sporadic parathyroid carcinoma for somatic *MEN1* mutations and found missense or frameshift mutations in 3/23 (13%).[24] These case reports and clinical research indicate that parathyroid carcinoma can occur, albeit rarely, in the context of MEN1 syndrome.

Conclusion

The observed interactions of menin, TGF-β and Smad3 provide a convincing explanation for the association between menin deficiency and the development of parathyroid adenomas and hyperparathyroidism. Menin appears to exert its tumor suppressor effect by an important facilitative role in the TGF-β signalling pathway which, if activated successfully, inhibits parathyroid cell proliferation and PTH production. While loss of menin is an integral part of the familial MEN1 syndrome, the far more common condition of sporadic parathyroid adenomas has been linked to acquired mutations in the menin gene in many cases. More recently, a tentative link between menin and parathyroid carcinoma has been demonstrated for the first time. However, it remains unclear why deficiency in menin leads to endocrine-specific tumors. Future research into the pathogenesis of parathyroid neoplasia is needed to clarify the place and timing of *MEN1* mutations in the progression from hyperplasia, to adenoma and possibly to carcinoma in a subset of cases.

References

1. Bilezikian JP, Silverberg SJ. Clinical practice. Asymptomatic primary hyperparathyroidism. N Engl J Med 2004; 350:1746-1751.
2. Yao K, Singer FR, Roth SI et al. Weight of normal parathyroid glands in patients with parathyroid adenomas. J Clin Endocrinol Metab 2004; 89:3208-3213.

3. Imanishi Y, Hosokawa Y, Yoshimoto K et al. Primary hyperparathyroidism caused by parathyroid-targeted overexpression of cyclin D1 in transgenic mice. J Clin Investig 2001; 107:1093-102.
4. Costa-Guda J, Tokura T, Roth SI et al. Mitochondrial DNA mutations in oxyphilic and chief cell parathyroid adenomas. BMC Endocr Disord 2007; 7:8.
5. Dotzenrath C, Teh BT, Farnebo F et al. Allelic loss of the retinoblastoma tumor suppressor gene: a marker for aggressive parathyroid tumors? J Clin Endocrinol Metab 1996; 81:3194-6.
6. Larsson C, Skogseid B, Oberg K et al. Multiple endocrine neoplasia type 1 gene maps to chromosome 11 and is lost in insulinoma. Nature (Lond) 1988; 332:85-7.
7. Brandi ML, Gagel RF, Angeli A et al. Guidelines for diagnosis and therapy of MEN type 1 and type 2. J Clin Endocrinol Metab 2001; 86:5658-71.
8. Heppner C, Kester MB, Agarwal SK et al. Somatic mutation of the MEN1 gene in parathyroid tumors. Nat Genet 1997; 16:375-8.
9. Farnebo F, Teh B, Kytola S et al. Alterations of the MEN1 gene in sporadic parathyroid tumors. J Clin Endocrinol Metab 1998; 83:2627-30.
10. Sowa H, Kaji H, Kitazawa R et al. Menin inactivation leads to loss of transforming growth factor β inhibition of parathyroid cell proliferation and parathyroid hormone secretion. Cancer Res 2004; 64(6):2222-8.
11. Lemos MC, Thakker RV. Multiple endocrine neoplasia type 1 (MEN1): analysis of 1336 mutations reported in the first decade following identification of the gene. Hum Mutat 2008; 29:22-32.
12. Naito J, Kaji H, Sowa H et al. Expression and functional analysis of menin in a multiple endocrine neoplasia type 1 (MEN1) patient with somatic loss of heterozygosity in chromosome 11q13 and un-identified germline mutation of the MEN1 gene. Endocrine 2006; 29:485-90.
13. Bhuiyan MM, Sato M, Murao K et al. Expression of menin in parathyroid tumors. J Clin Endocrinol Metab 2000; 85:2615-9.
14. Ross S, Hill CS. How the Smads regulate transcription. Int J Biochem Cell Biol 2008; 40:383-408
15. Kaji H, Canaff L, Lebrun JJ et al. Inactivation of menin, a Smad3-interacting protein, blocks transforming growth factor type β signalling. Proc Natl Acad Sci USA 2001; 98:3837-42.
16. Shattuck TM, Costa J, Bernstein M et al. Mutational analysis of Smad3, a candidate tumor suppressor implicated in TGF-β and menin pathways, in parathyroid adenomas and enteropancreatic endocrine tumors. J Clin Endocrinol Metab 2002; 87:3911-4.
17. Subramaniam N, Leong GM, Cock TA et al. Cross-talk between 1,25-dihydroxyvitamin D3 and transforming growth factor-β signalling requires binding of VDR and Smad3 proteins to their cognate DNA recognition elements. J Biol Chem 2001; 276:15741-6.
18. Kremer R, Bolivar I, Goltzman D et al. Influence of calcium and 1,25-dihydroxycholecalciferol on proliferation and proto-oncogene expression in primary cultures of bovine parathyroid cells. Endocrinology 1989; 125:935-41.
19. Arnold A, Brown MF, Urena P et al. Monoclonality of parathyroid tumors in chronic renal failure and in primary parathyroid hyperplasia. J Clin Investig 1995; 95:2047-53.
20. Malachi TD, Zevi U, Gafter A et al. DNA repair and recovery of RNA synthesis in uremic patients. Kidney Int 1993; 44:385-389.
21. Shan L, Nakamura Y, Murakami M et al. Clonal emergence in uremic parathyroid hyperplasia is not related to MEN1 gene abnormality Jpn. J Cancer Res 1999; 90:965-9.
22. Tahara H, Imanishi Y, Yamada T et al. Rare somatic inactivation of the multiple endocrine neoplasia type 1 gene in secondary hyperparathyroidism of uremia. J Clin Endocrinol Metab 2000; 85:4113-7.
23. Imanishi Y, Tahara H, Palanisamy N et al. Clonal chromosomal defects in the molecular pathogenesis of refractory hyperparathyroidism in uremia. J Am Soc Nephrol 2002; 13:1490-8.
24. Haven CJ, van Puijenbroek M, Tan MH et al. Identification of MEN1 and HRPT2 somatic mutations in paraffin-embedded (sporadic) parathyroid carcinomas. Clin Endocrinol (Oxf) 2007; 67:370-6.
25. Agha A, Carpenter R, Bhattacharya S et al. Parathyroid carcinoma in multiple endocrine neoplasia type 1 (MEN1) syndrome: two case reports of an unrecognised entity. J Endocrinol Invest 2007; 30:145-9.
26. Dionisi S, Minisola S, Pepe J et al. Concurrent parathyroid adenomas and carcinoma in the setting of multiple endocrine neoplasia type 1: presentation as hypercalcemic crisis. Mayo Clin Proc 2002; 77:866-9.

CHAPTER 9

Role of Menin in Neuroendocrine Tumorigenesis

Terry C. Lairmore* and Herbert Chen

Abstract

The menin protein encoded by the *MEN1* tumor suppressor gene is ubiquitously expressed and highly conserved evolutionarily. The combination of findings from current in vitro and in vivo studies has not yielded a comprehensive understanding of the mechanisms of menin's tumor suppressor activity or the specific role for menin in endocrine tumorigenesis, although its diverse interactions suggest possible pivotal roles in transcriptional regulation, DNA processing and repair and cytoskeletal integrity. This manuscript summarizes recent research findings including studies of global gene expression in MEN 1-associated neuroendocrine tumors and pivotal changes in intracellular signaling pathways associated with neuroendocrine tumorigenesis. Finally, the clinical applications provided by the understanding of the effects of *MEN1* gene mutations on neuroendocrine tumor development in patients with this familial cancer syndrome are discussed.

Introduction

Expression of the menin mRNA transcript can be demonstrated in most endocrine and nonendocrine tissues. Despite this ubiquitous expression, comparison of the menin protein sequence to available databases reveals no significant homology to other known protein families. The murine *Men1* gene demonstrates 98% homology[1,2] to the human gene sequence. Knockout of both *Men1* alleles in mice results in embryonic lethality,[3] suggesting that menin may have a broader role in the regulation of cell growth that is not limited to the endocrine tissues affected in patients with MEN 1 syndrome. Heterozygous $Men1^{+/-}$ mice demonstrate somatic loss of the wild type *Men1* allele in tumors[3] and develop a pattern of endocrine tumor formation similar to those observed in human MEN 1 syndrome. Menin is predominately a nuclear protein[4] that binds to JunD, a member of the AP-1 transcription factor family, and represses JunD mediated transcription.[5,6] In addition, menin has been shown to physically interact with a diverse variety of other proteins including transcription factors, DNA processing factors, DNA repair proteins and cytoskeletal proteins (Smad3, NF-κ-β, nm23, Pem, FANCD2, RPA2, ASK and others).[7-13] The combination of findings from all current studies has not yielded a clear picture of the mechanisms of menin's tumor suppressor activity or the specific role for menin in endocrine tumorigenesis, although its diverse interactions suggest possible pivotal roles in transcriptional regulation, DNA processing and repair and cytoskeletal integrity.

Previous studies have not focused specifically on neuroendocrine cells. However in other in vitro systems, some of the effects of menin have been elucidated. Overexpression of menin has been shown to diminish the tumorigenic phenotype of Ras-transformed NIH-3T3 cells, consistent with

*Corresponding Author: Terry C. Lairmore—Division of Surgical Oncology, Scott and White Memorial Hospital Clinic, Texas A&M University System Health Sciences Center College of Medicine, 2401 S. 31st Street, Temple, Texas, 76508, USA. Email: tlairmore@swmail.sw.org

SuperMEN1: Pituitary, Parathyroid and Pancreas, edited by Katalin Balogh and Attila Patocs. ©2009 Landes Bioscience and Springer Science+Business Media.

its putative tumor suppressor function.[14] In addition, studies have suggested a possible role for menin in repressing telomerase activity in somatic cells, perhaps explaining in part its tumor suppressor properties.[15] Menin has most recently been shown to regulate transcription in differentiated cells by associating with and modulating the histone methyltransferase activity of a nuclear protein complex to activate specific gene expression, including the cyclin-dependent kinase (CDK) inhibitors $p27^{Kip1}$ and $p18^{Ink4c}$,[16-18] as well as other cell cycle regulators (reviewed in detail by Lairmore 19).

Recent work has suggested that menin facilitates transcription of cell cycle regulators essential for normal endocrine cell growth control by promoting histone modifications within specific gene promoters. This suggests that menin may mediate its tumor suppressor action by regulating histone methylation in promoters of *HOX* genes and/or $p18^{Ink4c}$, $p27^{Kip1}$ and possibly other CDK inhibitors. The MLL (mixed-lineage leukemia) protein is a histone methyltransferase mutated in subsets of acute leukemia. Recent evidence has been provided that the menin tumor suppressor protein is an essential oncogenic cofactor for MLL-associated leukemogenesis.[16-18] In neuroendocrine cells, wild-type menin interacts with MLL to promote expression of antiproliferative (*p18, p27*) CDK inhibitors, possibly representing a central role in menin's tumor suppressor activity. Using genome-wide chromatin immunoprecipitation coupled with microarray analysis, it has also been recently demonstrated that menin frequently colocalizes with a protein complex that modifies chromatin structure and may also bind to many other promoters by an alternative mechanism.[20] Allelic loss of the murine *Men1* tumor suppressor in vitro in mouse embryonic fibroblasts accelerates cell cycle G_0/G_1 to S-phase transition and in vivo in a model in which floxed *Men1* alleles can be excised in a temporally controlled manner, directly enhances pancreatic islet cell proliferation.[21] Nonetheless, the comprehensive interactions of menin as a tumor suppressor gene and its specific roles in tumorigenesis are complex and have not been completely elucidated to date. This recent work has shed some light on potential critical roles of menin in neuroendocrine tumorigenesis.

Global Gene Expression in Normal Islet Cells versus MEN 1-Associated Neuroendocrine Tumors

Our workgroup performed a global gene expression analysis of 8 islet cell tumors arising in 6 patients with the MEN1 syndrome and compared the expressed gene levels with those obtained from 4 normal islet cell preparations.[22] This study represents the first analysis of global gene expression in MEN1-associated islet cell tumors. We used the Affymetrix U95Av2 chip, and a subset of 11 differentially expressed genes were validated by quantitative RT-RCR. Hierarchical clustering using all data separated the neoplastic group from the normal islets and within the tumor group those of the same functional type were mostly clustered together.

There were 193 genes differentially expressed (forty-five increased and 148 decreased) by at least 2-fold ($p <= 0.005$) in tumors relative to the normal islets. One hundred and four of the genes could be classified as being involved in cell growth, cell death, or signal transduction. In addition, the clustering analysis revealed 19 apoptosis-related genes that were under-expressed in tumors suggesting that these genes may play crucial roles in tumorigenesis in this syndrome. We identified a number of genes that are attractive candidates for further investigation into the mechanisms by which menin loss causes tumors in pancreatic islets. Of particular interest are the following: FGF9 which may stimulate the growth of prostrate cancer, brain cancer and endometrium; and IER3 (IEX-1), PHLDA2 (TSSC3), IAPP (amylin) and SST, all involved in apoptosis. IER3 (IEX-1) is regulated by several transcription factors and may have positive or negative effects on cell growth and apoptosis depending upon the cell-specific context.[23] Several studies have shown that it can be a promoter of apoptosis.[24] PHLDA2 (TSSC3) is an imprinted gene homologous to the murine TDAG51 apoptosis-related gene[25] and may be involved in human brain tumors.[26] IAPP (amylin) is a gene that has contrasting activities and has been associated with experimental diabetes in rodents.[27] Amylin deposits were increased in islet of patients with gastrectomy-induced islet atrophy.[28] On the other hand, exposure of rat embryonic islets to amylin results in β cell proliferation.[29] In contrast, amylin has been shown to induce apoptosis in rat and human insulinoma cells in vitro.[30,31]

Maitra et al[32] conducted a study that in many ways was similar to ours. They compared gene expression profiles of a series of sporadic pancreatic endocrine tumors with isolated normal islets, using the Affymetrix U133A chip. There was no overlap in genes they identified (having a three-fold or greater difference in expression) with the genes we identified (having a two-fold or greater difference in expression). This is quite surprising and if we don't take into account the different experimental settings, it might suggest that sporadically arising tumors have a quite different gene expression pattern than tumors arising as a result of menin loss of function.

A direct relationship between loss of menin function and any of the genes identified by this study[22] could not be established by this methodology. However, there were some genes, which because of their association with growth or apoptosis are of special interest. The general suppression of apoptosis related genes noted in our study (Fig. 1) has been highlighted by the recent study of Schnepp et al,[33] who showed that the loss of menin it was followed by suppression of apoptosis in murine embryonic fibroblast through a caspase-8 mechanism.

Recently, other studies of global gene expression in pancreatic islets have been performed.[32,34-36] Cardozo et al[37] have used microarrays to look for NF-κB dependent genes in primary cultures of rat pancreatic islets. Shalev et al[38] have measured global gene expression in purified human islets in tissue culture under high and low glucose concentrations. They noted that the TGFβ superfamily member prostate derived factor (PDF) was down regulated by 10-fold in the presence of glucose, whereas other TGFβ superfamily members were up regulated. In the current study, none of the

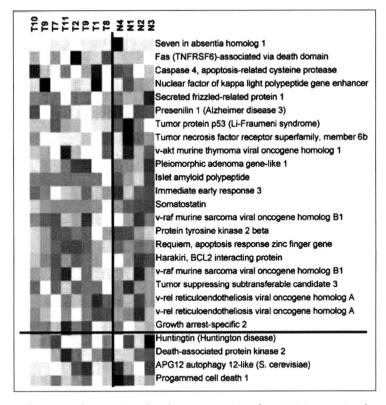

Figure 1. Clustering of apoptosis-related gene expression from MEN1-associated neuroendocrine tumors (T) and normal (N) islet cells by microarray analysis. Pink indicates strong, white indicates moderate and blue indicates weak expression. (Reproduced with permission from: Dilley WG et al. Mol Cancer 2005; 4(1):9;[22]). A color version of this image is available at www.landesbioscience.com/curie.

TGFβ superfamily members were significantly different between neoplastic and normal cells. Scearce et al[35] have used a pancrease-specific micro-chip, the PanChip to analyze gene expression patterns in embryonic day 14.5 to adult mice. Only a few specific genes were noted and none of them had human homologs.

Taking these data together, a diverse variety of genes have been shown to be differentially expressed in neuroendocrine tumors associated to MEN1 syndrome, as compared with normal human islet cells. Further investigations into the mechanisms by which menin loss causes tumors in pancreatic islets are needed.

Signaling Pathways in Neuroendocrine Tumors

Many signaling pathways, such as the phosphatidyl-inositol 3-kinase (PI3K)/Akt, mitogen activated protein kinases (MAPKs), (Fig. 2) and Notch1/Hairy Enhancer of Split-1 (HES-1)/ achaete-scute complex like-1 (ASCL1) signaling pathway (Fig. 3), have been recently shown to play important roles in regulating the growth of neuroendocrine tumors (NETs).[39-47] These pathways have been most extensively studied in carcinoid neuroendocrine cells.

ASCL1

Data from Chen and colleagues and others suggest that ASCL1 may play a key role in the development of NETs. NETs express high levels of ASCL1. In vivo abolition of ASCL1 in transgenic knockout mice led to the failed development of pulmonary NE cells, a lack of thyroid C-cells and a 50% reduction in adrenal chromaffin cell population.[48,49] These results clearly indicated that ASCL1 is required for the development of NE cells in the body including C-cells, adrenal chromaffin cells and pulmonary endocrine cells, the precursor cells for medullary thyroid cancer (MTC), pheochromocytoma and small cell lung cancer (SCLC), respectively.[47] We and others have characterized the expression of ASCL1 in several human cancer cell lines and tumors and found that ASCL1 is, as expected, highly expressed in NETs such as MTC, SCLC, carcinoids and pheochromocytoma, where as ASCL1 is absent in nonNET.[47,50-52] Therefore, inhibition of ASCL1 expression may be an important way to suppress NET growth.

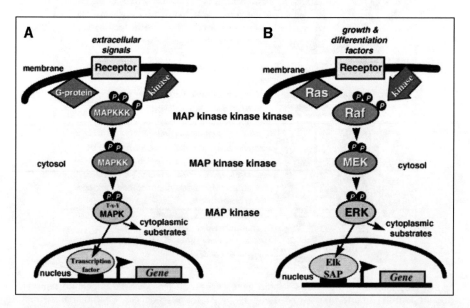

Figure 2. Schematic representation of the structure of the MAPK pathways. (Reproduced with permission from: Kolch W. Biochem J 2000; 351:289-305[73]).

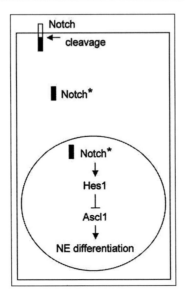

Figure 3. Schematic overview of inhibition of neuroendocrine differentiation in GI carcinoid tumors by Notch signaling. Upon ligand binding, the Notch receptor undergoes sequential internal cleavage, generating an activated intracytoplasmic domain (Notch*). Activated (cleaved) Notch* translocates to the nucleus and forms a transcriptionally-active complex, which activates the transcription of effector genes, such as Hes1. The repressor HES1 inhibits the expression of the proendocrine transcription factor gene Ascl1, resulting in the inhibition of NE differentiation. (Reproduced from: Nakakura EK et al. J Clin Endocrinol Metab 2005; 90(7):4350-6;[43] with permission from the Endocrine Society).

Notch1

The pathways that regulate ASCL1 expression have been well characterized. The Notch signaling pathway negatively regulates ASCL1 during Drosophila and mammalian development. Ligand activated Notch1 protein translocates to nucleus and partners with the CBF1 complex and acts as a transcriptional activator for various genes including HES-1, a transcriptional repressor of ASCL1. Interestingly, recent studies have shown that Notch1 signaling is very minimal or absent in NETs.[40,41,43,53,54] This could be the reason why we see high-level expression of ASCL1 protein in these tumors. Interestingly, we observed the absence of active Notch1 protein in pancreatic carcinoid BON cells suggesting that the Notch signaling pathway is inactive in carcinoids. Transient expression of active, Notch1 via adenoviral vector in BON cells resulted in growth suppression and significant reduction in NET markers such as serotonin, CgA, synaptophysin and ASCL1 confirming the tumor suppressor role of Notch1 signaling in carcinoid.[43] Further, it was shown that the reduction in serotonin is at the level of transcription of tryptophan hydroxylase 1 mRNA suggesting that Notch1 signaling regulates tryptophan hydroxylase 1, a rate-limiting enzyme in serotonin biosynthesis.[43] In addition, stable expression of a Notch1 fusion protein in BON cells also resulted in high levels of functional Notch1 that led to an increase in the level of HES-1. Increase in the level of HES-1 significantly reduced the level of ASCL1 protein. Similar to transient adenoviral Notch1 activation, the stable expression of Notch1 in BON cells also caused reductions in the levels of serotonin, CgA, NSE and synaptophysin.[40] However, the exact mechanisms by which growth and marker reduction remain unclear. Over expression of doxycycline inducible HES-1 in pulmonary carcinoid cells resulted in a dose dependent growth reduction as well as ASCL1 suppression. Interestingly, moderate growth reduction was observed with over expression of HES-1 in carcinoid cells.[41] This indicated that there could be additional factor(s) involved in Notch1 signaling pathway mediated growth suppression. These results demonstrate that Notch1 pathway components are intact in carcinoid cells and that these cells are capable of responding to Notch1 signaling. Importantly, these NET cells lack Notch1 activation at baseline. Therefore, identification of compound(s) that activates endogenous Notch1 in carcinoids should be exploited. This might result in clinical applications in the treatment of patients with carcinoid disease. Recently, we have shown that histone deacteylase (HDAC) inhibitors upregulate Notch1 in NETs and inhibit tumor growth.[55,56] Clinical trials with these agents are currently ongoing.

Raf-1

Ras regulates multiple signaling pathways of which the best understood is the Ras/Raf/mitogen-activated extracellular protein kinase (MEK)/extracellular signal-regulated kinase (ERK) pathway. Ras and raf are proto-oncogenes and expression of these genes activates signaling pathways, which in turn control cellular growth. Therefore, the ras/raf signaling pathway has been recognized as an important process in cancer biology. Despite several findings and new insights on this signaling pathway, the role of Raf in cancer cells remains controversial yet interesting. Recently, we have shown that activation of raf-1 pathway in MTC and other NETs by expression of estradiol inducible estrogen receptor fused with catalytic domain of raf-1 fusion protein led to complete suppression of ASCL1 mRNA and protein.[44,45,52,57] Decrease in the level of ASCL1 protein correlated with reduction in calcitonin and CgA. Furthermore, raf-1 activation in MTC cells led to a significant growth suppression. Further, it has been shown that growth inhibition by raf-1 activation in MTC-TT cell line induces an autocrine-paracrine protein, leukemia inhibitory factor (LIF) and this alone could mediate differentiation and cell growth inhibition.[58] Furthermore, we have recently shown that activation of raf-1 pathway in these cells lead to inactivation of GSK-3β by phosphorylation at ser9. We also showed that inactivation of the GSK-3β alone resulted in differentiation and cell growth inhibition.[56] These findings are interesting because raf-1 activation not only activates its own raf-1/MEK/ERK pathway but also cross talk with other pathways, which in turn could possibly regulate growth. Currently, clinical trials with GSK-3β inhibitors are underway for NETs including carcinoid, pancreatic islet cell cancer and MTC.

Treatment of Neuroendocrine Tumors Based on Molecular Genetic Diagnosis

Diverse genetic changes including nonsense, missense, frameshift and splice mutations, as well as large genomic deletions of the *MEN1* gene (reviewed by Schussheim et al[59]), have been reported. These mutations may occur anywhere within the coding sequence or the exon-intron junctions of the gene.[60] Two-thirds of the reported mutations of the *MEN1* gene result in truncation of the C-terminal portion of the menin protein. Recognized patters of genotype-phenotype relationships have not been established for MEN 1, although phenotypic variants (isolated hyperparathyroidism, or predominant prolactinomas) have been described.[61,62]

Genetic testing for MEN 1 is currently available in selected centers as a standard diagnostic test. Formal genetic counseling and informed consent including disclosures relevant to privacy of medical information and the potential impact of the genetic information on treatment are essential to a comprehensive program of genetic testing.

The optimal timing and most appropriate operation to perform for NETs of the pancreas and duodenum in patients with MEN 1 remains controversial. To recommend the appropriate surgical therapy, it is critical to understand the natural history of small, potentially benign or nonfunctional tumors. This understanding must be interpreted in the context of the risks of early and/or repeat major pancreatic interventions. These operations carry a significant risk of morbidity and even mortality. It is difficult to advocate routine or early pancreatic exploration in young otherwise healthy patients for small nonfunctional tumors, which are potentially clinically insignificant. Nevertheless, these tumors have a malignant potential and a delay in diagnosis and effective treatment carries the risk of spread to local or distant sites. It is obviously desirable to intervene early to prevent malignant dissemination, while minimizing morbidity and mortality (from either cancer or surgery). Surgical treatment decisions are made more difficult by the lack of evidence to support a relationship between size of the tumor and risk of regional lymph node or distant metastases.[63,64]

The opposite ends of the spectrum for surgical treatment of the neuroendocrine tumors that develop in association with MEN 1 include early, aggressive surgical intervention when the patient's serum tumor markers first become elevated (even without radiographically detectable tumors),[65-67] or operation only for tumors exceeding approximately 1.0 cm in size on radiographic imaging or demonstrating hormone hyperfunction.[68-71] The malignant potential of these neoplasms is clear and up to 50% of patients eventually develop regional lymph node or distant metastases.[63,68] Many groups now recommend early operation and excision of these tumors to prevent malignant progression.[63,68,70,72] A recent study reported improved overall survival in patients undergoing operation, especially in younger patients with localized tumors and in those with hormonally functional tumors.[63] The operative strategy for NET in patients with MEN 1 must be aimed at excision of grossly evident tumors with preservation of pancreatic exocrine and endocrine function with the safest operation that is effective.

References

1. Guru SC, Crabtree JS, Brown KD et al. Isolation, genomic organization and expression analysis of Men1, the murine homolog of the MEN1 gene. Mammalian Genome 1999; 10(6):592-6.
2. Stewart C, Parente F, Piehl F et al. Characterization of the mouse Men1 gene and its expression during development. Oncogene 1998; 17:2485-93.
3. Crabtree JS, Scacheri PC, Ward JM et al. A mouse model of multiple endocrine neoplasia, type 1, develops multiple endocrine tumors. Proceedings of the National Academy of Sciences of the USA 2001; 98(3):1118-23.
4. Guru SC, Goldsmith PK, Burns AL et al. Menin, the product of the MEN1 gene, is a nuclear protein. Proceedings of the National Academy of Sciences of the USA 1998; 95:1630-4.
5. Agarwal SK, Guru SC, Heppner C et al. Menin interacts with the AP1 transcription factor JunD and represses JunD-activated transcription. Cell 1999; 96:143-52.
6. Gobl AE, Berg M, Lopez-Egido JR et al. Menin represses JunD-activated transcription by a histone deacetylase-dependent mechanism. Biochim Biophys Acta 1999; 1447(1):51-6.
7. Heppner C, Bilimoria KY, Agarwal SK et al. The tumor suppressor protein menin interacts with NF-kappaB proteins and inhibits NF-kappaB-mediated transactivation. Oncogene 2001; 20(36):4917-25.
8. Kaji H, Canaff L, Lebrun JJ et al. Inactivation of menin, a Smad3-interacting protein, blocks transforming growth factor type β signaling. Proceedings of the National Academy of Sciences of the USA 2001; 98(7):3837-42.
9. Lemmens IH, Forsberg L, Pannett AA et al. Menin interacts directly with the homeobox-containing protein Pem. Biochem Biophys Res Commun 2001; 286(2):426-31.
10. Ohkura N, Kishi M, Tsukada T et al. Menin, a gene product responsible for multiple endocrine neoplasia type 1, interacts with the putative tumor metastasis protein nm23. Biochim Biophys Res Commun 2001; 282(5):1206-10.
11. Sukhodolets KE, Hickman AB, Agarwal SK et al. The 32-Kilodalton subunit of replication protein A interacts with menin, the product of the MEN1 tumor suppressor gene. Molecular and Cellular Biology 2003; 23:493-509.
12. Agarwal SK, Kennedy PA, Scacheri PC et al. Menin molecular interactions: insights into normal functions and tumorigenesis. Horm Metab Res 2005; 37(6):369-74.
13. Schnepp RW, Hou Z, Wang H et al. Functional interaction between tumor suppressor menin and activator of S-phase kinase. Cancer Res 2004; 64(18):6791-6.
14. Kim YS, Burns AL, Goldsmith PK et al. Stable overexpression of MEN1 suppresses tumorigenicity of RAS. Oncogene 1999; 18(43):5936-42.
15. Elledge SJ, Lin S-Y. Multiple tumor suppressor pathways negatively regulate telomerase. Cell 2003; 113:881-9.
16. Chen YX, Yan J, Keeshan K et al. The tumor suppressor menin regulates hematopoiesis and myeloid transformation by influencing Hox gene expression. Proc Natl Acad Sci USA 2006; 103(4):1018-23.
17. Yokoyama A, Somervaille TC, Smith KS et al. The menin tumor suppressor protein is an essential oncogenic cofactor for MLL-associated leukemogenesis. Cell 2005; 123(2):207-18.
18. Yokoyama A, Wang Z, Wysocka J et al. Leukemia proto-oncoprotein MLL forms a SET1-like histone methyltransferase complex with menin to regulate Hox gene expression. Mol Cell Biol 2004; 24(13):5639-49.
19. Lairmore TC, Moley JF. The multiple endocrine neoplasia syndromes. In: Townsend CM, ed. Sabiston Textbook of Surgery: The Biological Basis of Modern Surgical Practice. 18th ed. Philadelphia: W.B. Saunders Company, 2008:

20. Scacheri PC, Davis S, Odom DT et al. Genome-wide analysis of menin binding provides insights into MEN1 tumorigenesis. PLoS Genet 2006; 2(4):e51.
21. Schnepp RW, Chen YX, Wang H et al. Mutation of tumor suppressor gene Men1 acutely enhances proliferation of pancreatic islet cells. Cancer Res 2006; 66(11):5707-15.
22. Dilley WG, Kalyanaraman S, Verma S et al. Global gene expression in neuroendocrine tumors from patients with the MEN1 syndrome. Mol Cancer 2005; 4(1):9.
23. Wu MX. Roles of the stress-induced gene IEX-1 in regulation of cell death and oncogenesis. Apoptosis 2003; 8(1):11-8.
24. Schnepp RW, Mao H, Sykes SM et al. Menin induces apoptosis in murine embryonic fibroblasts. J Biol Chem 2004; Mar 12;279(11):10685-91. Epub 2003 Dec 18.
25. Lee MP, Feinberg AP. Genomic imprinting of a human apoptosis gene homologue, TSSC3. Cancer Res 1998; 58(5):1052-6.
26. Muller S, van den Boom D, Zirkel D et al. Retention of imprinting of the human apoptosis-related gene TSSC3 in human brain tumors. Hum Mol Genet 2000; 9(5):757-63.
27. Gebre-Medhin S, Olofsson C, Mulder H. Islet amyloid polypeptide in the islets of Langerhans: friend or foe? Diabetologia 2000; 43(6):687-95.
28. Itoh H, Takei K. Immunohistochemical and statistical studies on the islets of Langerhans pancreas in autopsied patients after gastrectomy. Human Pathology 2000; 31(11):1368-76.
29. Karlsson E, Sandler S. Islet amyloid polypeptide promotes β-cell proliferation in neonatal rat pancreatic islets. Diabetologia 2001; 44(8):1015-8.
30. Rumora L, Hadzija M, Barisic K et al. Amylin-induced cytotoxicity is associated with activation of caspase-3 and MAP kinases. Biological Chemistry 2002; 383(11):1751-8.
31. Zhang S, Liu J, MacGibbon G et al. Increased expression and activation of c-Jun contributes to human amylin-induced apoptosis in pancreatic islet β-cells. Journal of Molecular Biology 2002; 324(2):271-85.
32. Maitra A, Hansel DE, Argani P et al. Global expression analysis of well-differentiated pancreatic endocrine neoplasms using oligonucleotide microarrays. Clin Cancer Res 2003; 9(16 Pt 1):5988-95.
33. Schnepp RW, Mao H, Sykes SM et al. Menin induces apoptosis in murine embryonic fibroblasts. J Biol Chem 2004; 279(11):10685-91.
34. Cardozo AK, Heimberg H, Heremans Y et al. A comprehensive analysis of cytokine-induced and nuclear factor-kappa B-dependent genes in primary rat pancreatic β-cells. J Biol Chem 2001; 276(52):48879-86.
35. Scearce LM, Brestelli JE, McWeeney SK et al. Functional genomics of the endocrine pancreas: the pancreas clone set and PancChip, new resources for diabetes research. Diabetes 2002; 51(7):1997-2004.
36. Shalev A, Pise-Masison CA, Radonovich M et al. Oligonucleotide microarray analysis of intact human pancreatic islets: identification of glucose-responsive genes and a highly regulated TGFβ signaling pathway. Endocrinology 2002; 143(9):3695-8.
37. Cardozo AK, Heimberg H, Heremans Y et al. A comprehensive analysis of cytokine-induced and nuclear factor-kappa B-dependent genes in primary rat pancreatic β-cells. Journal of Biological Chemistry 2001; 276(52):48879-86.
38. Shalev A, Pise-Masison CA, Radonovich M et al. Oligonucleotide microarray analysis of intact human pancreatic islets: identification of glucose-responsive genes and a highly regulated TGFβ signaling pathway. Endocrinology 2002; 143(9):3695-8.
39. Kunnimalaiyaan M, Chen H. The Raf-1 pathway: a molecular target for treatment of select neuroendocrine tumors? Anticancer Drugs 2006; 17(2):139-42.
40. Kunnimalaiyaan M, Traeger K, Chen H. Conservation of the Notch1 signaling pathway in gastrointestinal carcinoid cells. Am J Physiol Gastrointest Liver Physiol 2005; 289(4):G636-42.
41. Kunnimalaiyaan M, Yan S, Wong F et al. Hairy Enhancer of Split-1 (HES-1), a Notch1 effector, inhibits the growth of carcinoid tumor cells. Surgery 2005; 138(6):1137-42; discussion 42.
42. Lal A, Chen H. Treatment of advanced carcinoid tumors. Curr Opin Oncol 2006; 18(1):9-15.
43. Nakakura EK, Sriuranpong VR, Kunnimalaiyaan M et al. Regulation of neuroendocrine differentiation in gastrointestinal carcinoid tumor cells by notch signaling. J Clin Endocrinol Metab 2005; 90(7):4350-6.
44. Sippel RS, Carpenter JE, Kunnimalaiyaan M et al. The role of human achaete-scute homolog-1 in medullary thyroid cancer cells. Surgery 2003; 134(6):866-71; discussion 71-3.
45. Sippel RS, Carpenter JE, Kunnimalaiyaan M et al. Raf-1 activation suppresses neuroendocrine marker and hormone levels in human gastrointestinal carcinoid cells. Am J Physiol Gastrointest Liver Physiol 2003; 285(2):G245-54.
46. Sippel RS, Chen H. Carcinoid tumors. Surg Oncol Clin N Am 2006; 15(3):463-78.
47. Van Gompel JJ, Sippel RS, Warner TF et al. Gastrointestinal carcinoid tumors: factors that predict outcome. World J Surg 2004; 28(4):387-92.
48. Borges M, Linnoila RI, van de Velde HJ et al. An achaete-scute homologue essential for neuroendocrine differentiation in the lung. Nature 1997; 386(6627):852-5.

49. Lanigan TM, DeRaad SK, Russo AF. Requirement of the MASH-1 transcription factor for neuroendocrine differentiation of thyroid C cells. J Neurobiol 1998; 34(2):126-34.
50. Chen H, Biel MA, Borges MW et al. Tissue-specific expression of human achaete-scute homologue-1 in neuroendocrine tumors: transcriptional regulation by dual inhibitory regions. Cell Growth Differ 1997; 8(6):677-86.
51. Chen H, Thiagalingam A, Chopra H et al. Conservation of the Drosophila lateral inhibition pathway in human lung cancer: a hairy-related protein (HES-1) directly represses achaete-scute homolog-1 expression. Proc Natl Acad Sci USA 1997; 94(10):5355-60.
52. Chen H, Udelsman R, Zeiger MA et al. Human achaete-scutehomolog-1 is highly expressed in a subset of neuroendocrine tumors. Oncology Reports 1997; 4:775-8.
53. Kunnimalaiyaan M, Vaccaro AM, Ndiaye MA et al. Overexpression of the NOTCH1 intracellular domain inhibits cell proliferation and alters the neuroendocrine phenotype of medullary thyroid cancer cells. J Biol Chem 2006; 281(52):39819-30.
54. Sriuranpong V, Borges MW, Ravi RK et al. Notch signaling induces cell cycle arrest in small cell lung cancer cells. Cancer Res 2001; 61(7):3200-5.
55. Greenblatt DY, Vaccaro AM, Jaskula-Sztul R et al. Valproic acid activates notch-1 signaling and regulates the neuroendocrine phenotype in carcinoid cancer cells. Oncologist 2007; 12(8):942-51.
56. Kunnimalaiyaan M, Vaccaro AM, Ndiaye MA et al. Inactivation of glycogen synthase kinase-3β, a downstream target of the raf-1 pathway, is associated with growth suppression in medullary thyroid cancer cells. Mol Cancer Ther 2007; 6(3):1151-8.
57. Chen H, Carson-Walter EB, Baylin SB et al. Differentiation of medullary thyroid cancer by C-Raf-1 silences expression of the neural transcription factor human achaete-scute homolog-1. Surgery 1996; 120(2):168-72; discussion 73.
58. Park JI, Strock CJ, Ball DW et al. The Ras/Raf/MEK/extracellular signal-regulated kinase pathway induces autocrine-paracrine growth inhibition via the leukemia inhibitory factor/JAK/STAT pathway. Mol Cell Biol 2003; 23(2):543-54.
59. Schussheim DH, Skarulis MC, Agarwal SK et al. Multiple endocrine neoplasia type 1: new clinical and basic findings. Trends in Endocrinology and Metabolism 2001; 12:173-8.
60. Mutch MG, Dilley WG, Sanjurjo F et al. Germline mutations in the multiple endocrine neoplasia type 1 gene: Evidence for frequent splicing defects. Human Mutation 1999; 13:175-85.
61. Kassem M, Kruse TA, Wong FK et al. Familial isolated hyperparathyroidism as a variant of multiple endocrine neoplasia type 1 in a large Danish pedigree. Journal of Clinical Endocrinology and Metabolism 2000; 85:165-7.
62. Olufemi SE, Green JS, Manickam P et al. Common ancestral mutation in the MEN1 gene is likely responsible for the prolactinoma variant of MEN 1 (MEN1 Burin) in four kindreds from Newfoundland. Human Mutation 1998; 11:264-9.
63. Kouvaraki MA, Shapiro SE, Cote GJ et al. Management of pancreatic endocrine tumors in multiple endocrine neoplasia type 1. World J Surg 2006; 30(5):643-53.
64. Lowney JK, Frisella MM, Lairmore TC et al. Pancreatic islet cell tumor metastasis in multiple endocrine neoplasia type 1: Correlation with primary tumor size. Surgery 1998; 124:1043-9.
65. Akerstrom G, Hessman O, Skogseid B. Timing and extent of surgery in symptomatic and asymptomatic neuroendocrine tumors of the pancreas in MEN 1. Langenbecks Arch Surg 2002; 386(8):558-69.
66. Skogseid B, Eriksson B, Lundqvist G et al. Multiple endocrine neoplasia type 1: A 10-year prospective screening study in four kindreds. Journal of Clinical Endocrinology and Metabolism 1991; 73:281-7.
67. Skogseid B, Oberg K, Eriksson B et al. Surgery for asymptomatic pancreatic lesion in multiple endocrine neoplasia type I. World J Surg 1996; 20(7):872-6; discussion 7.
68. Lairmore TC, Chen VY, DeBenedetti MK et al. Duodenopancreatic resections in patients with multiple endocrine neoplasia type 1. Annals of Surgery 2000; 231:909-18.
69. Thompson NW. Current concepts in the surgical management of multiple endocrine neoplasia type 1 pancreatic-duodenal disease. Results in the treatment of 40 patients with Zollinger-Ellison syndrome, hypoglycaemia or both. Journal of Internal Medicine 1998; 243:495-500.
70. Bartsch DK, Fendrich V, Langer P et al. Outcome of duodenopancreatic resections in patients with multiple endocrine neoplasia type 1. Ann Surg 2005; 242(6):757-64, discussion 64-6.
71. Thompson NW. Management of pancreatic endocrine tumors in patients with multiple endocrine neoplasia type 1. Surg Oncol Clin N Am 1998; 7(4):881-91.
72. Tonelli F, Fratini G, Nesi G et al. Pancreatectomy in multiple endocrine neoplasia type 1-related gastrinomas and pancreatic endocrine neoplasias. Ann Surg 2006; 244(1):61-70.
73. Kolch W. Meaningful relationships: the regulation of the Ras/Raf/MEK/ERK pathway by protein interactions. Biochem J 2000; 351:289-305.

Chapter 10

Adrenal Tumors in MEN1 Syndrome and the Role of Menin in Adrenal Tumorigenesis

Attila Patocs,* Katalin Balogh and Karoly Racz

Introduction

Most of the adrenal tumors are benign adrenocortical adenoma (AA) and pheochromocytomas (Pheo) originating from the adrenal medulla, but rarely malignant adrenocortical carcinomas (ACC) can be also found. Adrenal tumors causing hormonal overproduction such as aldosterone-producing and cortisol-producing tumors are also rare, whereas nonhyperfunctioning adenomas occur more frequently.[1] During the last decades an extensive use of advanced imaging techniques (computer tomography, magnetic resonance imaging, endoscopic ultrasound) has led to an increased incidence of accidentally discovered adrenal masses, i.e., incidentalomas.[2-4] The prevalence of incidentalomas is up to 9% of all autopsy cases. The majority of these tumors are hormonally inactive and are of adrenocortical origin, but pheochromocytomas and hormonally active adrenocortical tumors associated with the development of Cushing's syndrome or primary aldosteronism can also be found in some patient.[1,4]

Several clinical studies provided compelling evidence that adrenal tumors are associated with MEN1 syndrome and that patients with MEN1 syndrome may develop the entire spectrum of adrenal tumors including nonhyperfunctioning adenomas, cortisol- and aldosterone-producing tumors, adrenocortical carcinomas and, rarely, pheochromocytomas. However, adrenal tumors are not included in the main diagnostic components of MEN1 syndrome and patients with adrenal tumors who have only one of the three components without family history of MEN1 usually do not have mutations of the *MEN1* gene. Perhaps more interestingly, the somatic genetic alterations detected in MEN1-associated adrenal tumors do not seem to support a role for *MEN1* gene similar to that presumably involved in the pathomechanism of MEN1-associated parathyroid, pituitary or pancreas neuroendocrine tumors.

Genetics of Adrenal Tumors

Hereditary adrenocortical tumors are rare, but up to 25-30% of pheochromocytomas are associated with hereditary syndromes.[5,6]

Hereditary Syndromes with Adrenal Involvement

Hereditary adrenocortical tumors represent only a few percent of all adrenal tumors. In Li-Fraumeni syndrome (LFS; OMIM 151623), germline mutation of the tumorsuppressor gene

*Corresponding Author: Attila Patocs—Hungarian Academy of Sciences Molecular Medicine Research Group and 2nd Department of Medicine, Faculty of Medicine, Semmelweis University, Central Isotope Laboratory Semmelweis University, 1088 Budapest, Szentkirályi u. 46, Hungary. Email: patatt@bel2.sote.hu

SuperMEN1: Pituitary, Parathyroid and Pancreas, edited by Katalin Balogh and Attila Patocs. ©2009 Landes Bioscience and Springer Science+Business Media.

Table 1. Genetics of adrenal tumors. Hereditary tumor syndromes associated with adrenal tumors

	Gene and Chromosomal Localization	Tumors and Other Manifestations
Li-Fraumeni syndrome	TP53 (17q13)	ACC, breast cancers, brain tumors, soft tissue sarcoma leukaemia
Multiple endocrine neoplasia Type 1 (MEN1)	MEN1 (11q13)	3P: Parathyroid, pituitary, pancreas tumors; adrenal cortical; adenoma, hyperplasia, rarely carcinoma
Carney complex (CNC)	PRKARIA (17q22-24) PDE11A (2p16)	PPNAD, cardiac myxomas, GH- and PRL-secreting tumors, thyroid tumors, testicular tumors, ovarian cysts, lentiginosis,
Beckwith-Wiedemann syndrome (BWS)	11p15 locus alterations IGF-II overexpression	Omphalocele, macroglossia, macrosomia, hemihypertrophy, Wilms' tumor, ACC
Congenital adrenal hyperplasia (CAH)	CYP21A2 (6p)	Adrenal hyperplasia
Glucocorticoid-remediable aldosteronism (GRA)	CYP11B1 (8q21) CYP11B2 (8q21)	Micronodular, homogeneous hyperplasia

ACC: adrenocortical cancer; ADE: adrenocortical benign adenoma; GH: growth hormone; PRL: prolactin; IGF: insulin-like growth factor; GRA: Glucocorticoid-remediable aldosteronism.

TP53 can be identified,[7,8] whereas in Beckwith-Wiedemann syndrome (BWS; OMIM130650) overexpression of the insulin-like growth factor-2 (IGF-2) can be detected.[5] Mutations of the protein kinase A regulatory subunit-1α (*PRKAR1A*) gene have been associated with Carney complex (CNC; OMIM 160980).[9,10] In addition, mutation of the gene encoding the phosphodiesterase 11A enzyme (*PDE11A*) has been reported recently in patients with micronodular adrenocortical hyperplasia.[11] Multiple endocrine neoplasia Type 1 (MEN1; OMIM 131100) syndrome has been associated with mutations of the *MEN1* gene.[12] Glucocorticoid remediable hyperaldosteronism (GRA) is caused by a chimeric gene containing the promoter region of the *CYP11B1* (OMIM 610613) gene and the coding sequence of the *CYP11B2* gene (OMIM 12408) (Table 1).[13] Congenital adrenal hyperplasia is caused by mutations of the *CYP21A2* (*CYP21B*; OMIM 201910).[14,15] Activating mutations of the alpha chain of the stimulatory G protein have been described in McCune-Albright syndrome (MAS; OMIM 174800), but in this case somatic mosaicism occurs.[16]

Somatic Genomics of Sporadic Adrenal Tumors

Different molecular biological techniques, such as comparative genomic hybridization (CGH) and microsatellite analysis have been used in genome-wide screening for the identification of additional loci involved in adrenal tumorigenesis. Using CGH, chromosomal alterations have been observed in 28-61% of adrenal adenomas. Hot-spots for allelic losses have been identified on chromosomes 1p, 2q, 11q, 17p, 22p and 22q and gains on chromosomes 4, 5, 12 and 19.[17] Kjellman et al (1999) screened a panel of 60 tumors (39 carcinomas and 21 adenomas) for loss of heterozygosity (LOH). The vast majority of LOH detected was in the carcinomas involving chromosomes 2, 4, 11 and 18; but little was found in the adenomas. The Carney complex (160980) and the *MEN1* loci on 2p16 and 11q13, respectively, were further studied in 27 (13 carcinomas and 14 adenomas) of the 60 tumors. A detailed analysis of the 2p16 region mapped a minimal area of overlapping deletions to a 1-cM region that was separate from the Carney complex locus. LOH for glycogen phosphorylase gene (PYGM, OMIM 608455) was detected in all 8 informative carcinomas and in 2 of the 14 adenomas. Of the cases analyzed in detail, 13 of the 27 adrenal tumors (11 carcinomas and 2 adenomas) showed LOH on chromosome 11 and these were selected for *MEN1* mutation analysis. In 6 cases a common polymorphism was found, but no mutation was detected. The authors concluded that LOH in 2p16 was strongly associated with the malignant phenotype. In addition, LOH in 11q13 occurred frequently in carcinomas, but it was not associated with *MEN1* mutations, suggesting the involvement of a different tumor suppressor gene on this chromosome.[18]

Studies using microsatellite markers have demonstrated high percentages of loss of heterozygosity (LOH)/allelic imbalance at region 11q13 (in 100% of cases)[19-21] and 2p16 (in 92% of cases)[18] in adrenal carcinomas. LOH of the 17p13 locus has been reported to be highly specific to malignant tumors[22] and to be of prognostic value for the recurrence of localized tumors.[23] Based on these findings LOH at 11q13 occurs in about 20% of sporadic adrenal tumors, mostly in benign adrenocortical adenomas and in up to 40% of patients from MEN1 kindreds.[17-24]

Unlike the LOH at 11q13 detected in adrenal tumors, somatic mutations of the *MEN1* gene are very rare. Two studies conducted mutation screening of the *MEN1* gene. Heppner et al (1999) found no mutations within the coding region of the *MEN1* gene in 33 tumors and cell lines.[20] Schulte et al (1999) studied 16 patients with sporadic adrenal adenomas (4 patients had incidentally discovered masses, 5 patients had primary aldosteronism, 6 patients had Cushing's syndrome and one patient had multinodular hyperplasia) and only one patient with hormonally inactive adrenal adenoma showed a heterozygous missense mutation (Thr552Ser).[21] Retention of heterozygosity for the *MEN1* locus at 11q13 was also observed by Skogseid et al (1992), who analysed adrenocortical lesions in 31 MEN1 patients. Of the 31 patients, 12 (37%) had adrenal enlargement, which was bilateral in 7 patients. Of the 12 adrenal lesions 11 were benign adenomas and all retained heterozygosity for the *MEN1* locus. One interesting clinical observation on the association between adrenal and pancreatic endocrine tumors has been reported. In a single adrenocortical carcinoma, loss heterozygosity for alleles at 17p, 13q, 11p and 11q has been identified, which is in agreement with reports in sporadic cases. Skogseid et al (1992) concluded that the pituitary-independent adrenocortical proliferation is not the manifestation of a primary lesion in MEN I but it may represent a secondary phenomenon, perhaps related to the pancreatic endocrine tumor.[25]

MEN1-Associated Adrenal Tumors

Individuals who have a germline inactivating mutation of the *MEN1* gene develop MEN1 syndrome. In accordance with Knudson's two-hit hypothesis, their germline *MEN1* mutation combines with acquired somatic mutations of the second copy of their *MEN1* gene. This leads to monoclonal expansion and multiple neoplasia arises in such organs as the pituitary, parathyroid glands and the endocrine pancreas in an autosomal dominant manner.[26,27]

Prevalence

In the first family described by Wermer peptic ulcer and tumors of the anterior pituitary gland, parathyroid glands and islets of Langerhans, adenomas of the thyroid and of the adrenal cortex, as well as lipomas were identified.[28,29] In the past 50 years the prevalence of adrenal lesions observed in *MEN1* mutation carriers varied between 8 to 73% (Table 2.).[25,30-37]

Clinical Features

Similar to the usual presentation of sporadic adrenocortical tumors, adrenal cortical adenomas found in MEN1 usually are hormonally inactive but in a proportion <10% cortisol-secreting tumors can be found. Primary aldosteronism has also been occasionally reported.[38,39] Adrenocortical carcinomas (ACC) has been described only in a few cases and pheochromocytomas may occur in less than 1% of MEN1 patients.[32,33,40]

Table 2. Prevalence of the adrenal tumor in multiple endocrine neoplasia Type 1 (MEN1) among the MEN1 mutation carriers

Country of Origin	Number of Patients with Adrenal Tumor	Number of Total Mutation Carriers	Prevalence	Reference
Finland	29	82	35%	Vierimaa O et al
UK	5	59	8%	Ellard S et al
Germany (a multicenter study)	38	258	15%	Machens A et al
Germany	21	38	55%	Waldmann J et al
France	15	62	24%	Giraud S et al
Sweden	12	33	37%	Skogseid B et al
Hungary	1	10	10%	Balogh K et al
Germany (2008, EUS)	36	49	73%*	Schaefer S et al

*including plump adrenals.

Diagnosis, Therapy and Follow-Up of Adrenal Tumors

Localization

Most adrenal tumors found in patients with MEN1 syndrome are small, benign, hormonally inactive adrenocortical adenomas. These tumors are mostly discovered by routine imaging techniques, i.e., ultrasonography, CT or endoscopic ultrasound (EUS).[2,37] The sensitivity of EUS in the detection of adrenal lesions in MEN1 is higher than that observed with CT. Schaefer et al reported a very high percentage (73%) of patients with adrenal involvement among *MEN1* gene mutation carriers.[37] However, we should keep in mind that in this particular study all the EUS were performed by a single investigator who was searching for adrenal lesions. In routine clinical practice this high sensitivity could be difficult to reproduce when different investigators are doing the EUS examinations.[37]

Pheochromocytoma does not represent a major MEN1 manifestation, however, it can be observed in <1% of patients. 131I-MIBG (meta-iodo-benzyl guanidine) scintigraphy is a highly specific imaging technique for the diagnosis of pheochromocytoma.[41,42]

Laboratory Diagnosis

Although adrenal lesions in patients with MEN1 syndrome are mostly hormonally inactive adenomas, hormone secretion should be excluded. Urinary cortisol, midnight serum and salivary cortisol and the low-dose dexamethasone suppression test can be applied for the diagnosis of Cushing's syndrome. The plasma renin/aldosterone ratio is used for screening primary aldosteronism and adrenal androgens should also be determined.

In case of clinical suspicion of pheochromocytoma, urinary catecholamine metabolites (metanephrine, normetanephrine, homovanillic acid, vanillylmandelic acid) and serum chromogranin A determinations should be performed.[43]

Therapy

The therapeutical procedures for patients with MEN1-associated adrenal tumors are similar to patients with the sporadic counterparts. All functioning adrenal tumors and nonfunctioning tumors larger than 4 cm with evidence or suspicion of malignancy should be surgically resected. Nonfunctioning adrenal tumors smaller than 4 cm should be evaluated using imaging techniques and hormone measurements. Based on CT scan lipid-rich and lipid-poor tumor can be identified. If lipid-rich tumor is observed, imaging should be repeated after 6 month, while in cases when lipid-poor tumor is observed repeat CT or MRI is indicated after 3 months. If enlargement occurs adrenalectomy should be considered. In addition, adrenalectomy should be considered when heterogeneity, irregular capsule, nodes or change in hormonal activity are observed.[44,45]

Follow-Up

The main manifestations, pituitary adenomas, parathyroid hyperplasias causing primary hyperparathyroidism and pancreatic endocrine tumors are the major prognostic factors for the long-time survival of patients with MEN1. Primary treatment and additional work-up for these conditions can be found in Chapter 1 of this book.

Long-term follow-up for detection of adrenal tumors in MEN1 mutation carriers includes imaging studies and, if necessary, hormone measurements. An extensive use of imaging techniques, especially EUS, may lead to an increased prevalence of adrenal lesions in MEN1 patients. Using EUS Schaefer et al examined the natural course and clinical relevance of small adrenal lesions in MEN1 patients. They found that during a two-years follow-up period small adrenal lesions (<3 cm) were constant in their morphology. However, further studies with large number of patients would be needed to evaluate the course of adrenal lesions in *MEN1* mutation carriers and the influence of adrenal alterations on morbidity and mortality of these patients.[37]

MEN1 Gene Mutation Screening in Patients with Adrenal Tumors: To Screen or Not?

It is well established that hyperparathyroidism occurs in almost all patients with genetically confirmed MEN1 syndrome. Using current genetic tests mutation of the *MEN1* gene can be identified in 75-77% of patients with clinically well defined MEN1 syndrome. Patients with the MEN1 phenotype in whom genetic tests fail to confirm the presence of *MEN1* gene mutation may have alterations the promoter or introns which are missed by current routine mutation screening methods.[31,36]

The predictive value of different manifestations of the MEN1 syndrome for positive *MEN1* gene mutation detection is of particular interest. It has been shown that the best predictors of a positive genetic test are the number of main MEN1-associated tumors and the family history. *MEN1* gene mutation screening in our patients with a MEN1-related state who had a high prevalence of adrenal tumors but only one of the three main components without family history of MEN1 indicated a low prevalence of *MEN1* gene mutations. This finding may suggest, that the presence of adrenal tumors has a low predictive value for a positive *MEN1* mutation screening.[36] The impact of other tumors, such as lipomas, foregut, thymic and bronchial carcinoids, ependymomas and various cutaneous lesions on the probability of a positive *MEN1* gene mutation screening has been also analysed. One prospective study conducted by Asgharian et al assessed the frequency and sensitivity/specificity of various cutaneous alterations for MEN1 in 110 consecutive patients with gastrinomas with or without MEN1 syndrome. Interestingly, the presence of more than 3 angiofibromas or any collagenoma had the highest sensitivity (75%) and specificity (95%) for a positive *MEN1* gene mutation testing. They concluded that this diagnostic criterion has a greater sensitivity for MEN1 than pituitary or adrenal disease and has a sensitivity comparable to hyperparathyroidism reported in some studies of patients with MEN1 with gastrinoma.[46]

Another possibility would be to asses the decrease of *MEN1* function in patients by direct analysis of menin expression using real-time PCR or Western blotting techniques to compare menin expression in sporadic and MEN1-associated adrenal tumors. Until recently, only a few studies assessed the menin expression in adrenal tumors. Shulte et al analyzed 14 patients with sporadic adrenal cancer and menin mRNA expression was found in all tumors. Additionally, heterozygosity for the R176Q (in one patient) and for the D418D (in 40% of patients) were identified. In another study,[21] Bhuiyan et al, analyzed 12 different sporadic adrenal tumor tissues using RT-PCR and Western blotting. Upregulation of menin in Cushing's syndrome and a decreased expression in aldosterone-producing adrenal adenoma were detected. The authors concluded that upregulation of menin expression in Cushing's syndrome may result in an altered cellular function and it may represent an early step in adrenal carcinogenesis.[47] However, no further evidence for or against this hypothesis was presented. Retention of heterozygosity of the MEN1 locus observed in adrenal tumors of patients with MEN1 syndrome also supports the hypothesis that loss of the *MEN1* gene function is not the major cause of MEN1-associated adrenal tumors.[20,21,26]

Comments and Conclusion

Nonfunctional enlargement of one or both adrenal glands is a common finding in patients with MEN1 syndrome. In the majority of cases these adrenal lesions are hormonally inactive benign adrenocortical adenomas, but rarely pheochromocytomas, aldosterone-producing and cortisol-producing tumors and even adrenocortical carcinomas can also be found. Adrenal tumors in patients with mutations of the *MEN1* gene are phenotypically undistinguishable from their sporadic counterparts. A high prevalence of adrenal tumors in MEN-1 patients with pancreatic neuroendocrine tumors has been considered as the consequence of overexpression of growth factors such as proinsulin or insulin which may play a role in the pathomechanism of these tumors. The absence of loss of heterozygosity of the *MEN1* gene MEN1-associated adrenal tumors may indicate that inactivation of the *MEN1* gene does not rule tumorigenesis in the adrenal gland.

References

1. Beuschlein F, Reincke M. Adrenocortical tumorigenesis. Ann N Y Acad Sci 2006; 1088:319-334.
2. Bovio S, Cataldi A, Reimondo G et al. Prevalence of adrenal incidentaloma in a contemporary computerized tomography series. J Endocrinol Invest 2006; 29:298-302.
3. Soon PS, McDonald KL, Robinson BG et al. Molecular markers and the pathogenesis of adrenocortical cancer. Oncologist 2008; 13(5):548-61.
4. Igaz P, Wiener Z, Szabó P et al. Functional genomics approaches for the study of sporadic adrenal tumor pathogenesis: clinical implications. J Steroid Biochem Mol Biol 2006; 101(2-3):87-96.
5. Koch CA, Pacak K, Chrousos G-P. The molecular pathogenesis of hereditary and sporadic adrenocortical and adrenomedullary tumors. J Clin Endocrinol Metab 2002; 87(12):5367-5384.
6. Neumann H-P-H, Bausch B, McWhinney S-R et al. C. Eng and Freiburg—Warsaw—Columbus pheochromocytoma study group, Germ-line mutations in nonsyndromic pheochromocytoma. New Engl J Med 2002; 346(19):1459-1466.
7. Hisada M, Garber JE, Fung CY et al. Multiple primary cancers in families with Li-Fraumeni syndrome. J Natl Cancer Inst 1998; 90:606-611.
8. Frebourg T, Barbier N, Yan YX et al. Germ-line p53 mutations in 15 families with Li-Fraumeni syndrome. Am J Hum Genet 1995; 56:608-615.
9. Carney JA, Hruska LS, Beauchamp GD et al. Dominant inheritance of the complex of myxomas, spotty pigmentation and endocrine overactivity. Mayo Clin Proc 1986; 61:165-172.
10. Kirschner LS, Carney JA, Pack SD et al. Mutations of the gene encoding the protein kinase A type I-alpha regulatory subunit in patients with the Carney complex. Nat Genet 2000; 26:89-92.
11. Horvath A, Boikos S, Giatzakis C et al. A genome-wide scan identifies mutations in the gene encoding phosphodiesterase 11A4 (PDE11A) in individuals with adrenocortical hyperplasia. Nat Genet 2006; 38(7):794-800.
12. Chandrasekharappa SC, Guru SC, Manickam P et al. Positional cloning of the gene for multiple endocrine neoplasia-type 1. Science 1997; 276:404-407.
13. Lifton RP, Dluhy RG, Powers M et al. A chimaeric 11 β-hydroxylase/aldosterone synthase gene causes glucocorticoid-remediable aldosteronism and human hypertension. Nature 1992; 355:262-265.
14. Speiser PW, White PC. Congenital adrenal hyperplasia. N Engl J Med 2003; 349:776-788.
15. Jaresch S, Kornely E, Kley HK et al. Adrenal incidentaloma and patients with homozygous or heterozygous congenital adrenal hyperplasia. J Clin Endocrinol Metab 1992; 74:685-689.
16. Bianco P, Riminucci M, Majolagbe A et al. Mutations of the GNAS1 gene, stromal cell dysfunction and osteomalacic changes in nonMcCune-Albright fibrous dysplasia of bone. J Bone Miner Res 2000; 15:120-128.
17. Sidhu S, Marsh DJ, Theodosopoulos G et al. Comparative genomic hybridization analysis of adrenocortical tumors. Journal of Clinical Endocrinology and Metabolism 2002; 87:3467-3474.
18. Kjellman M, Roshani L, Teh BT et al. Genotyping of adrenocortical tumors: Very frequent deletions of the MEN1 locus in 11q13 and of a 1-centimorgan region in 2p16. J Clin Endocrinol Metab 1999; 84:730-735.
19. Kjellman M, Kallioniemi OP, Karhu R et al. Genetic aberrations in adrenocortical tumors detected using comparative genomic hybridization correlate with tumor size and malignancy. Cancer Res 1996; 56:4219-4223.
20. Heppner C, Reincke M, Agarwal SK et al. MEN1 gene analysis in sporadic adrenocortical neoplasms. Journal of Clinical Endocrinology and Metabolism 1999; 84:216-219.

21. Schulte KM, Mengel M, Heinze M et al. Complete sequencing and messenger ribonucleic acid expression analysis of the MEN1 gene in adrenal cancer. Journal of Clinical Endocrinology and Metabolism 2000; 85:441-448.
22. Yano T, Linehan M, Anglard P et al. Genetic changes in human adrenocortical carcinomas. Journal of the National Cancer Institute 1989; 81:518-523.
23. Gicquel C, Bertagna X, Gaston V et al. Molecular markers and long-term recurrences in a large cohort of patients with sporadic adrenocortical tumors. Cancer Research 2001; 61:6762-6767.
24. Gicquel C, Bertagna X, Schneid H et al. Rearrangements at the 11p15 locus and overexpression of insulin-like growth factor-II gene in sporadic adrenocortical tumors. J Clin Endocrinol Metab 1994; 78:1444-1453.
25. Skogseid B, Larsson C, Lindgren PG et al. Clinical and genetic features of adrenocortical lesions in multiple endocrine neoplasia type 1. J Clin Endocrinol Metab 1992; 75:76-81.
26. Marx SJ, Agarwal SK, Kester MB et al. Germline and somatic mutation of the gene for multiple endocrine neoplasia type 1 (MEN-1). J Intern Med 1998; 243:447-453.
27. Dong Q, Debelenko LV, Chandrasekharappa SC et al. Loss of heterozygosity at 11q13: analysis of pituitary tumors, lung carcinoids, lipomas and other uncommon tumors with familial multiple endocrine neoplasia type 1. J Clin Endocrinol Metab 1997; 82:1416-1420.
28. Paul Wermer. Genetic aspects of adenomatosis of endocrine glands. The American Journal of Medicine 1954; 16(3):363-371.
29. Paul Wermer. Endocrine adenomatosis and peptic ulcer in a large kindred: Inherited multiple tumors and mosaic pleiotropism in man. The American Journal of Medicine 1963; 35(2):205-212.
30. Vierimaa O, Ebeling TM, Kytölä S et al. Multiple endocrine neoplasia type 1 in Northern Finland; clinical features and genotype phenotype correlation. Eur J Endocrinol 2007; 157(3):285-94.
31. Ellard S, Hattersley AT, Brewer CM et al. Detection of an MEN1 gene mutation depends on clinical features and supports current referral criteria for diagnostic molecular genetic testing. Clin Endocrinol (Oxf) 2005; 62(2):169-75.
32. Machens A, Schaaf L, Karges W et al. Age-related penetrance of endocrine tumours in multiple endocrine neoplasia type 1 (MEN1): a multicentre study of 258 gene carriers. Clin Endocrinol (Oxf) 2007; 67(4):613-22.
33. Waldmann J, Bartsch DK, Kann PH et al. Adrenal involvement in multiple endocrine neoplasia type 1: results of 7 years prospective screening. Langenbecks Arch Surg 2007; 392(4):437-43.
34. Langer P, Cupisti K, Bartsch DK et al. Adrenal involvement in multiple endocrine neoplasia type 1. World J Surg 2002; 26(8):891-6.
35. Giraud S, Zhang CX, Serova-Sinilnikova O et al. Germ-line mutation analysis in patients with multiple endocrine neoplasia type 1 and related disorders. Am J Hum Genet 1998; 63(2):455-67.
36. Balogh K, Hunyady L, Patocs A et al. MEN1 gene mutations in Hungarian patients with multiple endocrine neoplasia type 1. Clin Endocrinol (Oxf) 2007; 67(5):727-34.
37. Schaefer S, Shipotko M, Meyer S et al. Natural course of small adrenal lesions in multiple endocrine neoplasia type 1: an endoscopic ultrasound imaging study. Eur J Endocrinol 2008; 158(5):699-704.
38. Beckers A, Abs R, Willems P et al. Aldosterone-secreting adrenal adenoma as part of multiple endocrine neoplasia type 1 (MEN1): loss of heterozygosity for polymorphic chromosome 11 deoxyribonucleic acid markers, including the MEN1 locus. J Clin Endocrinol Metab 1992; 75:564-570.
39. Iida A, Blake K, Tunny T et al. Allelic losses on chromosome band 11q13 in aldosterone-producing adrenal tumors. Gene Chromosome Cancer 1995; 12:73-75.
40. Dohna M, Reincke M, Mincheva A et al. Adrenocortical carcinoma is characterized by a high frequency of chromosomal gains and high-level amplifications. Genes Chromosomes Cancer 2000; 28:145-152.
41. Furuta N, Kiyota H, Yoshigoe F et al. Diagnosis of pheochromocytoma using 123I-compared with 131I-metaiodobenzylguanidine scintigraphy. Int J Urol 1999; 6:119-124.
42. d'Herbomez M, Gouze V, Huglo D et al. Chromogranin A assay and 131I-MIBG scintigraphy for diagnosis and follow-up of pheochromocytoma. J Nucl Med 2001; 42:993-997.
43. Lenders JWM, Pacak K, Walther MM et al. Biochemical diagnosis of pheochromocytoma: which is the best test? JAMA 2002; 287:1427-1434.
44. NCCN Clinical Practice Guidelines in Oncology, V.I.2007, www.nccn.org
45. Brandi ML, Gagel RF, Angeli A et al. Guidelines for diagnosis and therapy of MEN type 1 and type 2. J Clin Endocrinol Metab 2001; 86:5658-5671.
46. Asgharian B, Turner ML, Gibril F et al. Cutaneous tumors in patients with multiple endocrine neoplasm type 1 (MEN1) and gastrinomas: prospective study of frequency and development of criteria with high sensitivity and specificity for MEN1. J Clin Endocrinol Metab 2004; 89(11):5328-36.
47. Bhuiyan MM, Sato M, Murao K et al. Differential expression of menin in various adrenal tumors. The role of menin in adrenal tumors. Cancer 2001; 92(6):1393-401.
48. Zwermann O, Beuschlein F, Mora P et al. Multiple endocrine neoplasia type 1 gene expression is normal in sporadic adrenocortical tumors. Eur J Endocrinol 2000; 142:689-695.

CHAPTER 11

Functional Studies of Menin through Genetic Manipulation of the *Men1* Homolog in Mice

Dheepa Balasubramanian and Peter C. Scacheri*

Abstract

To investigate the physiological role of menin, the protein product of the *MEN1* gene, several groups have utilized gene targeting strategies to delete one or both copies of the mouse homolog *Men1*. Mice that are homozygous null for *Men1* die during embryogenesis. Heterozygous *Men1* mice are viable and develop many of the same types of tumors as humans with MEN1. In addition to conventional knockouts of *Men1*, tissue-specific elimination of menin using cre-lox has been achieved in pancreatic β cells, anterior pituitary, parathyroid, liver, neural crest and bone marrow, with varying results that are dependent on cell context. In this chapter, we compare the phenotypes of the different conventional *Men1* knockouts, detail the similarities and differences between *Men1* pathogenesis in mice and humans and highlight results from recent crossbreeding studies between *Men1* mutants and mice with null mutations in genes within the retinoblastoma pathway, including $p18^{Inc4c}$, $p27^{Kip1}$ and *Rb*. In addition, we discuss not only how the *Men1* mutants have shed light on the role of menin in endocrine tumor suppression, but also how *Men1* mutant mice have helped uncover previously unrecognized roles for menin in development, leukemogenesis and gestational diabetes.

Introduction

Multiple Endocrine Neoplasia Type I (MEN1) is an autosomal dominant disease generally characterized by the occurrence of multiple tumors in multiple endocrine organs. The causative gene for MEN1 behaves like a classic tumor suppressor and strictly adheres to Knudson's two-hit model first proposed for retinoblastoma in 1971.[1] Patients with familial MEN1 inherit the first *MEN1* mutation through the germline, followed by the somatic loss of the second *MEN1* allele through chromosome loss, breakage, duplication, mitotic recombination or point mutation. Tumors from MEN1 patients consistently show loss of heterozygosity (LOH) at the *MEN1* locus, consistent with the notion that tumors arise after somatic loss of the wild type *MEN1* allele. The *MEN1* gene is also frequently mutated in endocrine tumors from patients with no family history of endocrine cancer, indicating that the *MEN1* gene is a major contributor to the development and maintenance of many nonhereditary endocrine tumors.[2-6]

The gene for MEN1 codes for menin, a 67kDa nuclear protein expressed in a wide variety of tissues and throughout embryogenesis. Menin bears no similarity to any known proteins and

*Corresponding Author: Peter C. Scacheri—Department of Genetics, Case Western Reserve University, Cleveland, OH, USA. Email: pxs183@case.edu

SuperMEN1: Pituitary, Parathyroid and Pancreas, edited by Katalin Balogh and Attila Patocs. ©2009 Landes Bioscience and Springer Science+Business Media.

contains no functional motifs, making it exceptionally challenging to elucidate its function. In spite of this, menin has been found to interact with more than twenty proteins.[7] The diverse functions of the menin partners suggest roles for menin in transcriptional regulation, DNA processing and repair, cytoskeletal organization and protein degradation. There is particularly strong evidence that menin functions as a coregulator of transcription. Specifically, independent studies have shown that menin interacts with a histone methyltransferase (HMT) complex containing MLL (mixed lineage leukemia), ASH2L (absent small, or homeotic), WDR5 (WD repeat containing protein 5) and RBBP5 (retinoblastoma binding protein 5).[8,9] The menin-HMT-associated complexes promote methylation of lysine 4 on histone H3 (H3K4me), an epigenetic mark that is often associated with transcriptionally active genes. Menin and HMT proteins colocalize to hundreds of gene promoters marked with H3K4me[10] and a subset of the menin-HMT target genes are misexpressed in *Men1*-null cells (including the potent cell cycle inhibitors $p27^{Kip1}$ and $p18^{Ink4c}$).[11,12] Menin can also bind to promoters independently of HMT proteins, suggesting that menin may also modulate transcription by an alternative coregulatory mechanism.[10]

In addition to the biochemical approaches described above, several groups have taken advantage of the in vivo study of mouse models, which utilizes gene targeting to delete one or both copies of the mouse *Men1* gene.[13-15] Human and mouse menin are 96.7% homologous,[16] and gratifyingly, the phenotype of mice that are heterozygous for *Men1* (*Men1*$^{+/-}$) is remarkably similar to that observed in humans with MEN1. Specifically, *Men1*$^{+/-}$ mice develop tumors of the pancreatic islets, pituitary, parathyroid and adrenal glands. Moreover, as in human cases of MEN1, tumors from heterozygous *Men1* mice all arise following somatic loss of the wild type allele. In addition to the conventional knockouts, several groups have successfully developed conditional, or tissue-specific knockouts of the *Men1* gene using the cre-lox system. To date, the complete inactivation of menin has been achieved in pancreatic β cells, parathyroid, hepatocytes, bone marrow and neural crest, with varying results dependent on cell context.[17-21] In this chapter, we review the phenotype of the *Men1* mouse mutants and discuss how these mice support menin's role as a tumor suppressor. In addition we will highlight studies that reveal previously unrecognized roles for menin in development, blood-related cancers and gestational diabetes.[21-23] Lastly, we will draw attention to recent crossbreeding studies which, together with biochemical studies, suggest that menin may mediate its tumor suppressor function in part by collaborating with p18 to restrain cell growth.[24,25]

Conventional *Men1* Mouse Models

Three different groups generated conventional mouse knockouts of *Men1* by disrupting the *Men1* mouse homolog through homologous recombination in embryonic stem (ES) cells.[13,15,20] Mice lacking both copies of the *Men1* gene (*Men1*$^{-/-}$) die around the time of mid-gestation at embryonic day (E) 11.5-13.5. *Men1* null embryos studied just prior to death are small in size and have accompanying neural, craniofacial, heart and liver malformations.[14,26] The results of these studies clearly demonstrate a critical role for menin in development of multiple organs, although the precise role of menin during embryogenesis and the underlying cause of the developmental defects remains largely unknown. One possibility is that the developmental defects are due to dysregulated expression of developmental genes that are normally coregulated by menin and its associated transcriptional proteins. The developmental homeobox genes are directly regulated by the menin-HMT complex and *Men1*$^{-/-}$ embryos show decreased expression of *Hoxc6* and *Hoxc8* when compared to wild type *Men1*$^{+/+}$ embryos.[8] Further studies are necessary to determine if the decreases in *Hox* gene expression have a direct affect on the developmental anomalies in the *Men1*$^{-/-}$ embryos. Presumably genes besides *Hox* are also coregulated by menin during embryogenesis and it will also be interesting to explore whether these genes might contribute to defects in *Men1*$^{-/-}$ embryos.

Mice lacking one copy of *Men1* (*Men1*$^{+/-}$) clearly develop many features of human MEN1. The most common types of tumors found in *Men1*$^{+/-}$ mice are insulinomas, prolactinomas, parathyroid adenomas, adrenal cortical tumors, lung tumors, testicular and ovarian tumors and thyroid adenomas (Table 1).[13-15] After adjusting for the difference in life expectancy, the age of onset in

Table I. Men1 conventional knockout mice

Region of Deletion; Strain	Reported Pathology Men1+/−	Comparison to Humans	Reference
Men1^{TSM}/+*, Men1 exons 3-8 (Men1^{Δ3-8}/+); C57/129SvEvTacFBR, C57BL/6	Die at E11.5-12.5, small in size, cranial and/or facial defects	Pancreatic islet hyperplasia or adenoma (28-83) (mostly insulin-secreting), parathyroid dysplasia or adenoma (24), anterior pituitary adenoma (26-43) (prolactin-secreting), adrenal cortical adenoma (20-43), pheochromocytoma (7), lung adenocarcinoma (22), Leydig cell tumor (22), thyroid abnormalities.**	Higher incidence of insulinoma and lower incidence of PTH tumors in Men1 mice. Gastrinomas and cutaneous tumors not observed in Men1 mice. Gonadal tumors not observed in humans with MEN1.	14,24
Men1 exon 3; 129/Ola, 129/Sv	Die at E11.5-E13.5; small in size nonclosure of neural tube, heart hypotrophy, altered liver organization	Pancreatic islet hyper/dysplasia (30) or adenoma (18) or carcinoma (44) (insulin, glucagon or gastrin-secreting), parathyroid adenoma (53), extrapancreatic adenoma (12), anterior pituitary adenoma (11) or carcinoma (15) (growth hormone- or prolactin-secreting), adrenal hyperplasia (12) or adenoma (21) or carcinoma (10), Leydig cell hyperplasia (30) or tumor (65), sex-cord stromal cell tumor (36), mammary gland carcinoma (6).***	Similar to Men1^{Δ3-8}/+, except for appearance of gastrinomas and mammary tumors in Men1 mice.	13,26
Men1 exon 2; C57/129	Not reported, but presumably similar to above	Pancreatic islet hyperplasia (17) or adenoma (mostly insulin-secreting) (79), anterior pituitary microadenoma or adenoma (78% females; 42% males), parathyroid gland hyperplasia (10) or adenoma (7), adrenal cortical microadenoma or tumor (8), sex cord stromal tumor (16), Leydig cell hyperplasia (13) or tumor (34), thyroid abnormalities (15-55).****	Similar to Men1^{Δ3-8}/+	15

Number in parentheses shows the percentage of tumors. *Men1^{TSM}/+ mice harbor a floxed PGK-neomycin cassette in an Men1 intron, creating a hypomorphic allele. **Incidences reported for Men1^{TSM}/+ or Men1^{Δ3-8}/+ mice ranging in age from 273-670 days. ***Total incidences reported for Men1+/− mice ranging in age from 270 to >540 days. ****Total incidences reported for Men1+/− mice ranging in age from 243-790 days. Tumor frequencies <5% are not reported.

mice (9-18 months) is comparable to humans. As seen in human tumors from MEN1 patients, tumors from $Men1^{+/-}$ mice show loss of the wild type $Men1$ allele.

There are notable differences in the incidence and type of tumors observed in humans and mice with MEN1. Whereas 90-99% of affected MEN1 patients develop parathyroid tumors, only 7-53% of $Men1$ mice develop parathyroid tumors. Cutaneous tumors in human MEN1 patients are not observed in $Men1^{+/-}$ mice. In addition, MEN1 patients develop pancreatic islet cell tumors 30-80% of the time and nearly half of these are gastrinomas, 10% are insulinomas and 10% are nonhormone secreting. By comparison, pancreatic islet cell tumors in affected mice are almost always insulinomas. Prior to developing insulinomas, the $Men1^{+/-}$ mice develop islet cell hyperplasia due to haploinsufficiency of $Men1$, a feature that is not observed in MEN1 patients. Lastly, gonadal tumors in $Men1^{+/-}$ mice are almost never seen in humans with MEN1.

The phenotypes of the different conventional $Men1$ knockout mouse models are quite similar, but differences between them do exist. Most notably, pancreatic islet and pituitary carcinomas, extra-pancreatic gastrinomas and a few growth-hormone secreting pituitary tumors are reported in one model[13] but not the other two.[14,15] The phenotypic variations may be due to subtle differences in the genetic backgrounds of the mice, which may contain genetic modifiers that contribute to the phenotype in unknown ways. At some point, studying these differences may lead to discovery of genetic contributors to phenotypic variability in patients with MEN1.

Conditional *Men1* Mouse Mutants

In designing the conventional *Men1* mouse knock out models, researchers artificially create the "first hit" by homologous recombination in ES cells. Mutation of the second *Men1* allele, or the "second hit", occurs naturally, similar to humans with germline mutations in *MEN1*. Because of this, the conventional models lend themselves to in vivo research of the natural progression of MEN1. However, the conventional models cannot be used to address the consequence of simultaneous loss of both *Men1* alleles in adult tissue, since complete absence of menin causes embryonic lethality. As a means to bypass the embryonic lethality, clever studies to manipulate the *Men1* gene in specific tissues have been designed. These conditional targeting strategies use cre-lox to selectively delete *Men1* at a specific time in tissues of interest. Specifically, a knock-in mouse with a "floxed" *Men1* allele is created by inserting loxP sites into the *Men1* locus by homologous recombination. *Men1* floxed mice are then bred to transgenic mice expressing cre recombinase from tissue specific promoters. In progeny that contain both the floxed *Men1* alleles and the cre transgene, cre recombinase binds to the loxP sites and catalyzes excision of the intervening sequence, resulting in inactivation of the endogenous *Men1* gene in the corresponding tissue. So far, this strategy has been used to address the consequences of homozygous deletion of *Men1* in several endocrine and nonendocrine tissues. The results are summarized in Table 2 and described below.

Table 2. Conditional Men1 knockout mice

Cre	Target Tissue	Phenotype	Reference
RIP-cre	Pancreatic β cell	Insulinomas, pituitary adenomas, high serum insulin, low blood glucose	17,18,20
ER-cre	Pancreatic β cell	Acute islet hyperplasia	27
PTH-cre	Parathyroid	Parathyroid neoplasms, hypercalcemia	29
Alb-cre	Liver	Normal liver, pancreatic insulinomas and glucagonomas	30
UBC9-cre-ERT2	Bone marrow	Reduced number of peripheral white blood cells	19
Wnt -cre, Pax3-cre	Neural crest	Perinatal death, cleft palate, rib patterning defects, craniofacial abnormalities	21

Pancreas and Pituitary

The most commonly studied tissue in *Men1* mutant mice has been pancreatic islets, which have the distinction of being more easily accessible and of higher quantity than other endocrine tissues associated with MEN1. By breeding floxed *Men1* mice to transgenic mice expressing cre from the rat insulin promoter (RIP-cre), several groups have inactivated menin in the pancreatic β cell.[17,18,20] The complete absence of menin in β cells leads to the formation of multiple insulinomas. Similar to humans and the conventional *Men1* knockout mice, the formation of insulinomas in the conditional knockout results in elevated serum insulin and decreased blood glucose levels. The incidence of insulinoma formation is high in the conditional knockouts and in some cases, the tumor burden gets to be such that the mice die from hypoglycemia if fasted for >12 hours. In addition to insulinomas, 18-56% of floxed *Men1* mice expressing RIP-cre develop large anterior pituitary adenomas. Although the pituitary tumors arise from leaky expression of cre recombinase in the pars distalis, they are histologically and functionally similar to those observed in human MEN1 patients.[18,20] There is also a striking sex bias in the mice, with females developing up to twice as many pituitary adenomas as males. The bias for pituitary adenoma formation in females is also reported in one of the conventional *Men1* models[15] and probably exists in the other two conventional models as well. The biological basis for this gender difference is not known.

Both *Men1* alleles are inactivated from a very early age in the β cells of the conditional model and it is therefore not surprising that islet tumorigenesis is accelerated. By breeding floxed *Men1* mice to mice expressing cre from an inducible estrogen receptor promoter (ER-cre) islet cell proliferation was determined to increase as early as 7 days following excision of *Men1*.[27] In the RIP-cre mice with floxed MEN1 alleles, large hyperplastic islets are seen as early as 4 weeks and tumors at 22-28 weeks, compared to 30-38 and 40-60 weeks in the conventional model. However, even with early loss of both copies of *Men1*, tumor appearance in the conditional knockout is considerably delayed. The delay in tumor formation clearly implies that loss of *Men1* alone is not sufficient for insulinoma formation and that additional somatic events are essential. $Men1^{-/-}$ insulinomas are capable of developing in the absence of chromosomal or microsatellite instability, suggesting that the additional somatic events required for tumor formation are probably subtle, occurring at either the nucleotide level or through epigenetic mechanisms.[28] As DNA sequencing costs decrease, it should become practical to identify the nucleotide changes associated with *Men1* tumorigenesis on a genome-wide scale.

Parathyroid

By breeding floxed *Men1* mice to transgenics expressing cre from the human parathyroid-gene promoter (PTH-cre), the *Men1* gene was inactivated in parathyroid glands.[29] None of the resulting mice developed frank parathyroid adenomas, although histological findings were consistent with parathyroid neoplasia and serum hypercalcemia at 7 months of age. Its rather surprising that even with early loss of both *Men1* alleles in the parathyroid, the PTH-specific knockouts are not as susceptible to developing PTH tumors as humans, or $Men1^{+/-}$ mice. The reasons for the disparities are not known, but the discrepancies in mice might be due to variations in genetic background.

Liver

To assess the effect of menin loss in an MEN1 nonsusceptible tissue, the *Men1* gene was excised in hepatocytes by crossing *Men1* floxed mice to mice expressing an albumin-promoter driven cre transgene.[30] Livers that were completely null for menin expression appeared entirely normal and remained tumor free. The mice did however develop pancreatic-islet tumors, which was attributed to leaky expression of cre. The data not only suggest that menin function is dispensable in the liver, but also that tissue-specificity in MEN1 is probably not determined by the likelihood of loss of the wild type *Men1* allele. Otherwise these mice would have bypassed this limiting step and developed multiple liver tumors.

Leukemic Precursor Cells and Normal Bone Marrow

Translocations involving MLL can create fusion proteins that lead to liquid cancers like acute lymphocytic leukemia and acute myelogenous leukemia. The fusion proteins behave as potent oncogenes by constitutively activating *HOX* genes, specifically *HOXA9* and its cofactor, *MEIS1*. This results in a failure of terminal differentiation of hematopoietic progenitors. Wild type MLL directly interacts with menin and importantly, abnormal (oncogenic) MLL-fusion proteins retain their ability to bind to menin.[9,22] The interaction between menin and the MLL fusion protein is required for oncogenic transformation of the myeloid cells. In MLL-transformed cells harboring floxed *Men1* alleles, conditional inactivation of menin suppresses the aberrant expression of *Hoxa9*.[9] Upon reduction of *Hoxa9*, the leukemic progenitor cells lose their oncogenic qualities and normal cell differentiation resumes, even in the presence of the abnormal fusion protein. Based on these findings, it was concluded that MLL-associated leukemogenesis requires the presence of menin. These conclusions were corroborated by a similar study in which proliferation of MLL-AF9 transformed myeloid cells was suppressed upon conditional inactivation of menin.[19] These studies present opportunities for new therapeutic targets, introducing the possibility for treating MLL-related leukemias through targeted removal of menin.

In addition to being involved in abnormal proliferation of blood cells in leukemia, *Men1* conditional knockout mice were used to show that menin is essential for normal blood cell development. Mice deficient for menin in bone marrow have reduced numbers of peripheral white blood cells.[19] Although this study supports a role for menin in hematopoiesis in mice, hematological abnormalities in humans with MEN1 are not reported.

Neural Crest

During development, cells derived from the neural crest form parts of the peripheral nervous system, parts of the face and head and the outflow tract of the heart. To investigate whether the craniofacial, neural and heart defects observed in *Men1*-null embryos are related to a function of menin in the neural crest, *Pax3*- and *Wnt1*-promoter driven cre transgenes were used to excise floxed *Men1* alleles in neural crest.[21] The resulting mice survived embryogenesis but died shortly after birth. In addition, cleft palate and other craniofacial bone abnormalities were observed. The data support a role for menin in palatogenesis and perinatal viability. Although a mechanism was not determined, the expression of p27 was reduced in *Men1*-null neural crest cells, suggesting that disturbances in genes directly targeted by menin might be a contributing factor. This is not the first time that a role for menin in bone formation has been suggested. Previous studies in mouse mesenchymal stem cells showed that menin, through regulated interactions with Smads and Runx2, promotes the formation of the osteoblasts.[31,32]

Menin Overexpression

During pregnancy, maternal β cell mass increases to meet physiological demands. Given the strong connection between reduced menin levels and islet cell proliferation, it was hypothesized that fluctuations in menin levels might modulate the adaptive β cell response in pregnancy. Consistent with this hypothesis, β cell menin levels drop at the onset of pregnancy in mice, accompanied by β cell proliferation and increased insulin production, both of which restore to normal after pregnancy.[23] Transgenic expression of menin in the β cell prevented expansion of the islets and similar to women with gestational diabetes, pregnant *Men1* transgenic mice developed hyperglycemia and had impaired glucose tolerance. The results of this study are particularly novel because they suggest that one of menin's normal functions is to respond to the body's changing physiological demands for insulin by controlling β cell proliferation. Although follow-up studies in human are necessary, the results also suggest that gestational diabetes might be related to defects in signaling pathways that control menin levels.

Crossbreeding Studies

p18 and p27 inhibit cell cycle progression by signaling through the retinoblastoma protein (Rb) (Fig. 1) and biochemical studies show that *p18* and *p27* are directly regulated by menin. Consistent with these findings, *p18* and *p27* are decreased in tissues derived from the *Men1* conventional and tissue-specific knockouts, including pancreatic islets, liver, fibroblasts and neural crest.[10,12,21] Mice that are homozygous null for either *p18* or *p27* develop gigantism and hyperplasia of

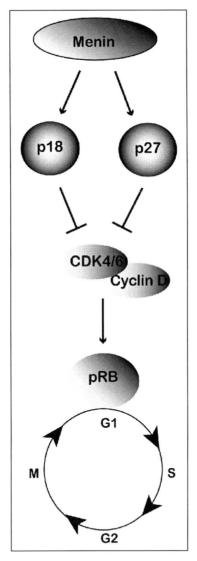

Figure 1. Menin is a positive regulator of p18 and p27, which inhibit cyclin dependent kinases CDK4/6. Phosphorylation of Rb by cyclin dependent kinases (CDK4/6 and cyclin D) initiates the transition from G1 to S-phase by activating transcription of S-phase promoting genes.

multiple organs by 2-3 weeks of age and pituitary tumors of the intermediate lobe by 10 months.[33-36] Interestingly, mice simultaneously lacking both $p18^{Ink4c}$ and $p27^{Kip1}$ develop features of MEN, including hyperplasia and/or tumors of the pituitary, adrenal glands, duodenum, stomach, testes, thyroid, parathyroid and endocrine pancreas.[37] In rats, germline mutations in $p27$ alone result in an MEN-like syndrome, with the affected animals displaying parathyroid adenomas, thyroid C-cell hyperplasia, pancreatic hyperplasia and pheochromacytomas.[38] Perturbations of other genes within the Rb pathway also yield phenotypes that are similar to those observed in the *Men1* mutant mice. For example, pancreatic β cell hyperpasia occurs in mice expressing a mutant form of cdk4 (CKD4-R24C) which is unresponsive to the inhibitory effects of INK4.[39] β-cell specific overexpression of cyclin D1 results in islet hyperplasia.[40] Human patients with inherited mutations in RB-1 develop retinoblastoma and osteosarcoma, but $Rb^{+/-}$ mice develop anterior and intermediate lobe pituitary tumors, metastatic thyroid carcinomas and pheochromocytomas.[37] These data raise the possibility that menin suppresses endocrine tumorigenesis in part by collaborating with key components of the Rb pathway to restrain cell cycle progression. This hypothesis has prompted crossbreeding studies between the *Men1* mutants and mice with null mutations in the *Rb* pathway, including $p18$, $p27$ and Rb.

Compared to single *Men1* mutants, $Men1^{+/-} p18^{-/-}$ compound mice show an increased incidence of anterior lobe pituitary tumors, insulinomas, parathyroid adenomas, adrenal cortical adenomas and lung tumors (Table 3).[24,25] In addition, many of these tumors develop faster than in *Men1* single mutants (3-12 months compared to 12-22 months in the compound mutants). Unlike tumors from *Men1* single mutants which all show LOH, tumors from $Men1^{+/-} p18^{-/-}$ mice retain the wild type Men1 allele, indicating that in the complete absence of p18, one-half dosage of menin is not sufficient to restrain tumor formation. Together with the biochemical data, the additive effects of the *Men1* and $p18$ mutations support the notion that the mechanism of tumor suppression by menin likely involves p18. Specifically, loss of menin reduces p18 expression, abrogating its inhibitory effect on cyclin dependent kinases and ultimately resulting in unrestrained cell growth.

Compared to single *Men1* mutants, neither $Men1^{+/-} p27^{-/-}$, or $Men1^{+/-} Rb^{+/-}$ compound mice show an appreciable increase in the incidence or rate of tumor formation (Table 3).[24,41,42] Although these data do not provide evidence for a functional collaboration between *Men1* and $p27$, or *Men1* and *Rb*, the data do not necessarily exclude the possibility that p27 and/or Rb are involved in MEN1 pathogenesis. p27 is a direct menin target whose expression is reduced in tissues from the single *Men1* mutants. If p27 function in the single *Men1* mutants is already maximally compromised, then complete elimination of p27 by genetic knockout would not necessarily enhance tumorigenesis in $Men1^{+/-} p27^{-/-}$ mice. As an alternative functional test, it might be worthwhile to breed $Men1^{+/-}$ mice to transgenics overexpressing p27. If increased p27 expression in $Men1^{+/-}$ mice delayed or prevented tumor formation, a role for p27 in MEN1 pathogenesis could be supported.

Conclusion

The *Men1* mutant mice have clearly helped further our understanding of menin function and MEN1 pathogenesis, but there is still much to be learned. For example, it is not known which of the more than twenty menin-interacting proteins are relevant to menin function. Crossbreeding the *Men1* mutants to mice that are genetically null for the menin partners may help sort out the roles of these proteins, if any, in MEN1 pathogenesis. It also remains to be elucidated whether any of the menin-interacting proteins are misexpressed or mislocalized in *Men1* null tissues. Given their MEN-related phenotypes, crossbreeding studies between *Men1* mice and *CDK4* and *cyclin D* mutants might also provide additional support for the involvement of Rb pathway members in MEN1 pathogenesis.

Lastly, its not clear why the loss of *Men1* gives rise to tumors in the endocrine organs and not many other tissues where menin is normally expressed. The data from the crossbreeding studies implicate p18 in MEN1 pathogenesis and biochemical studies support a role for p27. However, p18 and p27 can not alone account for the bias towards endocrine tumor formation

Table 3. *Tumor formation in* p18⁻/⁻, p27⁻/⁻, Rb⁺/⁻, Men1⁺/⁻, p18⁻/⁻; Men1⁺/⁻, p27⁻/⁻; Men1⁺/⁻, Rb⁺/⁻; Men1⁺/⁻ and p18⁻/⁻; p27⁻/⁻ mice

Phenotype	p18⁻/⁻	p27⁻/⁻	Rb⁺/⁻	Men1⁺/⁻	p18⁻/⁻; Men1⁺/⁻	p27⁻/⁻; Men1⁺/⁻*	Rb⁺/⁻; Men1⁺/⁻	p18⁻/⁻; p27⁻/⁻
Pituitary Tumor								
Intermediate lobe	+++	++	+++		+++	++	+++	+++
Anterior lobe			++	++/+++	+++	+	++	
Both				+++	+			
Pancreas								
Hyper/dysplasia	+	+	++	+/++		+	++	+
Insulinoma	+			++	+++	+		
Thyroid								
Hyperplasia/adenoma	+	+	+++	+	+++	+	+++	+++
C-cell carcinoma			+++				+++	
Parathyroid								
Adenoma				+/++	+++		++	
Testis								
Leydig cell tumor	+			+	++		+	
Adrenal								
Cortical adenoma				+/++	++		+	
Medullary								
hyperplasia/adenoma	+	+++			++	+		+
Pheochromocytoma	+	++	++	+			+	++
Lung				+			+++	
Adenoma	+					+		
Adenocarcinoma	+	+			++	++		
Neuroendocrine carcinoma					+	++		

Tumor incidences reported for *p18⁻/⁻, p27⁻/⁻, Men1⁺/⁻* and *p18⁻/⁻; Men1⁺/⁻* mice at 12-22 months. *Tumor incidence at 3-12 months. + = 1-33%, ++ = 34-66%, +++ = 67-100%.

in MEN1, since their expression is reduced in *Men1* null nonendocrine tissues that do not form tumors, including liver, fibroblasts and neural crest.[10,12,21] Given the evidence supporting the role of menin as a transcriptional coregulator, we hypothesize that the specific bias for endocrine tumor formation in MEN1 results from dysregulated expression of distinct genes that are targeted by menin and its cofactors in the endocrine tissues. One candidate, identified through a genome-wide screen for menin target genes, is *Hlxb9*, a developmentally programmed transcription factor that is overexpressed in islets in the absence of menin.[10] The absence of menin probably affects other endocrine-specific targets in addition to *Hlxb9* and its clear that the *Men1* mouse mutants will serve as a valuable resource for functionally testing these candidates as they are uncovered.

Acknowledgements

We are grateful to the members of the scientific community for their contributions to the study of MEN1 and menin function. We also acknowledge the support of the National Cancer Institute (KCA103843A, awarded to PCS).

References

1. Knudson AG Jr. Mutation and cancer: Statistical study of retinoblastoma. Proc Natl Acad Sci USA 1971; 68:820-3.
2. Debelenko LV, Brambilla E, Agarwal SK et al. Identification of MEN1 gene mutations in sporadic carcinoid tumors of the lung. Hum Mol Genet 1997; 6:2285-90.
3. Farnebo F, Teh BT, Kytola S et al. Alterations of the MEN1 gene in sporadic parathyroid tumors. J Clin Endocrinol Metab 1998; 83:2627-30.
4. Heppner C, Kester MB, Agarwal SK et al. Somatic mutation of the MEN1 gene in parathyroid tumors. Nat Genet 1997; 16:375-8.
5. Zhuang Z, Ezzat SZ, Vortmeyer AO et al. Mutations of the MEN1 tumor suppressor gene in pituitary tumors. Cancer Res 1997; 57:5446-51.
6. Zhuang Z, Vortmeyer AO, Pack S et al. Somatic mutations of the MEN1 tumor suppressor gene in sporadic gastrinomas and insulinomas. Cancer Res 1997; 57:4682-6.
7. Agarwal SK, Kennedy PA, Scacheri PC et al. Menin molecular interactions: insights into normal functions and tumorigenesis. Horm Metab Res 2005; 37:369-74.
8. Hughes CM, Rozenblatt-Rosen O, Milne TA et al. Menin associates with a trithorax family histone methyltransferase complex and with the hoxc8 locus. Mol Cell 2004; 13:587-97.
9. Yokoyama A, Wang Z, Wysocka J et al. Leukemia proto-oncoprotein MLL forms a SET1-like histone methyltransferase complex with menin to regulate hox gene expression. Mol Cell Biol 2004; 24:5639-49.
10. Scacheri PC, Davis S, Odom DT et al. Genome-wide analysis of menin binding provides insights into MEN1 tumorigenesis. PLoS Genet 2006; 2:e51.
11. Karnik SK, Hughes CM, Gu X et al. Menin regulates pancreatic islet growth by promoting histone methylation and expression of genes encoding p27Kip1 and p18INK4c. Proc Natl Acad Sci USA 2005; 102:14659-64.
12. Milne TA, Hughes CM, Lloyd R et al. Menin and MLL cooperatively regulate expression of cyclin-dependent kinase inhibitors. Proc Natl Acad Sci USA 2005; 102:749-54.
13. Bertolino P, Tong WM, Galendo D et al. Heterozygous Men1 mutant mice develop a range of endocrine tumors mimicking multiple endocrine neoplasia type 1. Mol Endocrinol 2003; 17:1880-92.
14. Crabtree JS, Scacheri PC, Ward JM et al. A mouse model of multiple endocrine neoplasia, type 1, develops multiple endocrine tumors. Proc Natl Acad Sci USA 2001; 98:1118-23.
15. Loffler KA, Biondi CA, Gartside M et al. Broad tumor spectrum in a mouse model of multiple endocrine neoplasia type 1. Int J Cancer 2007; 120:259-67.
16. Guru SC, Crabtree JS, Brown KD et al. Isolation, genomic organization and expression analysis of Men1, the murine homolog of the MEN1 gene. Mamm Genome 1999; 10:592-6.
17. Bertolino P, Tong WM, Herrera PL et al. Pancreatic β-cell-specific ablation of the multiple endocrine neoplasia type 1 (MEN1) gene causes full penetrance of insulinoma development in mice. Cancer Res 2003; 63:4836-41.
18. Biondi CA, Gartside MG, Waring P et al. Conditional inactivation of the Men1 gene leads to pancreatic and pituitary tumorigenesis but does not affect normal development of these tissues. Mol Cell Biol 2004; 24:3125-3131.
19. Chen YX, Yan J, Keeshan K et al. The tumor suppressor menin regulates hematopoiesis and myeloid transformation by influencing Hox gene expression. Proc Natl Acad Sci USA 2006; 103:1018-23.
20. Crabtree JS, Scacheri PC, Ward JM et al. Of mice and MEN1: Insulinomas in a conditional mouse knockout. Mol Cell Biol 2003; 23:6075-85.
19. Engleka KA, Wu M, Zhang M et al. Menin is required in cranial neural crest for palatogenesis and perinatal viability. Dev Biol 2007; 311:524-37.
22. Yokoyama A, Somervaille TC, Smith KS et al. The menin tumor suppressor protein is an essential oncogenic cofactor for MLL-associated leukemogenesis. Cell 2005; 123:207-18.
23. Karnik SK, Chen H, McLean GW et al. Menin controls growth of pancreatic β-cells in pregnant mice and promotes gestational diabetes mellitus. Science 2007; 318:806-9.
24. Bai F, Pei XH, Nishikawa T et al. p18Ink4c, but not p27Kip1, collaborates with Men1 to suppress neuroendocrine organ tumors. Mol Cell Biol 2007; 27:1495-504.
25. Pei XH, Bai F, Smith MD et al. p18Ink4c collaborates with Men1 to constrain lung stem cell expansion and suppress nonsmall-cell lung cancers. Cancer Res 2007; 67:3162-70.

26. Bertolino P, Radovanovic I, Casse H et al. Genetic ablation of the tumor suppressor menin causes lethality at mid-gestation with defects in multiple organs. Mech Dev 2003; 120:549-60.
27. Schnepp RW, Chen YX, Wang H et al. Mutation of tumor suppressor gene Men1 acutely enhances proliferation of pancreatic islet cells. Cancer Res 2006; 66:5707-15.
28. Scacheri PC, Kennedy AL, Chin K et al. Pancreatic insulinomas in multiple endocrine neoplasia, type I knockout mice can develop in the absence of chromosome instability or microsatellite instability. Cancer Res 2004; 64:7039-44.
29. Libutti SK, Crabtree JS, Lorang D et al. Parathyroid gland-specific deletion of the mouse Men1 gene results in parathyroid neoplasia and hypercalcemic hyperparathyroidism. Cancer Res 2003; 63:8022-8.
30. Scacheri PC, Crabtree JS, Kennedy AL et al. Homozygous loss of menin is well tolerated in liver, a tissue not affected in MEN1. Mamm Genome 2004; 15:872-7.
31. Sowa H, Kaji H, Canaff L et al. Inactivation of menin, the product of the multiple endocrine neoplasia type 1 gene, inhibits the commitment of multipotential mesenchymal stem cells into the osteoblast lineage. J Biol Chem 2003; 278:21058-69.
32. Sowa H, Kaji H, Hendy GN et al. Menin is required for bone morphogenetic protein 2- and transforming growth factor β-regulated osteoblastic differentiation through interaction with Smads and Runx2. J Biol Chem 2004; 279:40267-75.
33. Fero ML, Rivkin M, Tasch M et al. A syndrome of multiorgan hyperplasia with features of gigantism, tumorigenesis and female sterility in p27(Kip1)-deficient mice. Cell 1996; 85:733-44.
34. Franklin DS, Godfrey VL, Lee H et al. CDK inhibitors p18(INK4c) and p27(Kip1) mediate two separate pathways to collaboratively suppress pituitary tumorigenesis. Genes Dev 1998; 12:2899-911.
35. Kiyokawa H, Kineman RD, Manova-Todorova KO et al. Enhanced growth of mice lacking the cyclin-dependent kinase inhibitor function of p27(Kip1). Cell 1996; 85:721-32.
36. Nakayama K, Ishida N, Shirane M et al. Mice lacking p27(Kip1) display increased body size, multiple organ hyperplasia, retinal dysplasia and pituitary tumors. Cell 1996; 85:707-20.
37. Franklin DS, Godfrey VL, O'Brien DA et al. Functional collaboration between different cyclin-dependent kinase inhibitors suppresses tumor growth with distinct tissue specificity. Mol Cell Biol 2000; 20:6147-58.
38. Pellegata NS, Quintanilla-Martinez L, Siggelkow H et al. Germ-line mutations in p27Kip1 cause a multiple endocrine neoplasia syndrome in rats and humans. Proc Natl Acad Sci USA 2006; 103:15558-63.
39. Rane SG, Dubus P, Mettus RV et al. Loss of Cdk4 expression causes insulin-deficient diabetes and Cdk4 activation results in β-islet cell hyperplasia. Nat Genet 1999; 22:44-52.
40. Zhang X, Gaspard JP, Mizukami Y et al. Overexpression of cyclin D1 in pancreatic β-cells in vivo results in islet hyperplasia without hypoglycemia. Diabetes 2005; 54:712-9.
41. Loffler KA, Biondi CA, Gartside MG et al. Lack of augmentation of tumor spectrum or severity in dual heterozygous Men1 and Rb1 knockout mice. Oncogene 2007; 26:4009-17.
42. Matoso A, Zhou Z, Hayama R et al. Cell lineage-specific interactions between Men1 and Rb in neuroendocrine neoplasia. Carcinogenesis 2008; 29:620-8.

INDEX

A

Acromegaly 3, 4, 6, 7, 9, 12
Activin 41, 69, 71-75
Adrenal tumor 3, 4, 7, 9, 10, 97-102
Adrenocortical tumor 3, 7, 9, 97, 99
Anterior pituitary 1, 3, 7, 17, 38, 41, 59, 60, 69-73, 75, 83, 84, 99, 105, 107, 109
Ataxia telangiectasia mutated gene (ATM) 22, 27, 30, 33, 43
ATM and Rad3-related kinase (ATR) 23, 27, 30, 33, 39, 43
Autosomal dominant 1, 2, 5, 17, 70, 80, 81, 99, 105

B

Bone morphogenetic protein (BMP) 41, 42, 59, 60-65, 72, 73

C

Cancer 2-5, 8, 12, 17, 20, 23, 38, 45, 55, 60, 71, 75, 80, 87-90, 92, 98, 101, 105, 106, 110, 114
Carcinoid tumor 6, 8, 9, 91
Cell cycle 17, 22, 23, 30, 31, 37-41, 43, 44, 59, 60, 70, 74, 88, 106, 111, 112
Cell growth inhibition 73-75, 92
CHES1 23, 31, 39, 45
Chromatin 17, 23, 28-31, 37, 43, 45, 47, 88
Chromatin remodeling 37, 45, 47
Cushing's syndrome 3, 4, 9, 97, 99-101
Cyclin dependent kinase inhibitor 37

D

DNA damage response 23, 27, 30
Drosophila melanogaster 23, 30, 33

E

Endocrine pancreas tumor 3, 4, 6-8, 11
Endocrine tumorigenesis 21, 87, 112
Evolution 19, 31, 37

F

Familial 1, 2, 5, 6, 9, 10, 22, 70, 79, 80, 85, 87, 105
Foregut carcinoid 1, 6, 7, 9, 10, 23

G

Gastrinoma 1, 3, 6-10, 12, 23, 101, 107, 108
Gene expression 29, 31, 47, 52, 53, 56, 60, 72, 74, 87-90, 106
Genetic homogeneity 17
Genetic testing 92
Germline mutation 2, 5, 9, 10, 21, 80, 82, 83, 97, 108, 112
Glucagonoma 1, 3, 7, 108
GTPase 28, 30, 33, 40

H

Heat shock protein 31, 65
Hematopoiesis 47, 51, 53, 54, 110
Hematopoietic stem cell 51, 53, 54, 56
Hereditary tumor syndrome 1, 98
Histone deacetylase 27, 29, 37, 39, 40, 45, 60
Histone methylation 40, 88
Histone methyltransferase 31, 37, 46, 47, 51, 52, 88, 106
Homeobox domain gene 37
Hypercortisolism 3, 6, 7
Hyperparathyroidism 1, 2, 5, 6, 11, 12, 21, 44, 47, 79-83, 85, 92, 101

I

Imaging 8-10, 12, 69, 93, 97, 100, 101
Incidentaloma 97
Insulinoma 3, 7, 8-10, 17, 88, 106-109, 112, 113
Intracellular signaling 87
Islet beta cell 47

J

JunD 29-31, 40, 41, 46, 59-61, 64, 65, 81, 87

L

Laboratory diagnosis 8, 100
Lactotrope cell 69, 73, 75
Leukemia 5, 40, 47, 51, 54, 56, 88, 92, 106, 110

M

Maintenance of genome stability 22
Menin 1, 17-23, 27-31, 33-47, 51-56, 59-66, 69, 70, 73-75, 79, 81-85, 87-90, 92, 97, 101, 105, 106, 108-114
Mixed lineage leukemia complex 47, 52-54
Mixed-lineage leukemia (MLL) 23, 40, 47, 51-56, 88, 106, 110
Monogenic disease 2, 17
Mouse model 22, 40, 106, 108
mSin3A 27, 29, 40, 46, 60
Multiple Endocrine Neoplasia Type I (MEN1) 1-4, 6-12, 17, 18, 20-23, 27-31, 33, 40, 41, 43, 47, 53, 59, 70, 73, 75, 79-85, 87-90, 92, 97-102, 105-110, 112-114
Mutation 1, 2, 5, 6, 9, 10, 12, 17, 18, 20-23, 27-31, 33, 38, 43, 45, 60, 70-72, 75, 79-83, 85, 87, 92, 97-102, 105, 108, 112

N

Neural crest 40, 41, 66, 105, 106, 108, 110, 111, 113
Neuroendocrine tumor 12, 87-90, 92, 93, 97, 102
NF-κB 89
NM23H1 30, 40
Nonendocrine manifestation 4-7
Nuclear localization signal 27, 28, 31, 33, 44
Nuclear matrix 23, 27, 30, 45

P

$p18^{Inc4c}$ 105
$p27^{Kip1}$ 37, 53, 88, 105, 106, 112
Parathyroid 1-4, 6, 8, 11, 17, 21-23, 38, 41, 44, 47, 53, 61, 70, 75, 79-85, 97-99, 101, 105-109, 112, 113
Phenocopy 6, 12
Pituitary 1-4, 6-10, 12, 17, 22, 38, 41, 43, 59, 60, 69-75, 81, 83, 84, 97-99, 101, 105-109, 112, 113

Pituitary adenoma 3, 6, 38, 69-72, 74, 75, 101, 107-109
Prolactin 3, 8-10, 12, 47, 69, 71-75, 98, 107
Prolactinoma 3, 4, 6, 7, 12, 21, 69-72, 82, 92, 106
Protein interaction 27, 30, 33, 34

R

Retinoblastoma Protein (Rb) 37-40, 44, 65, 105, 111-113

S

Signal transduction 23, 71, 73-75, 88
Smad 37, 38, 41, 43, 60, 61, 69, 71, 73-75, 79, 81, 83, 84, 110
Somatostatin analogue 8, 11
Somatostatin receptor scintigraphy 8, 12
Sporadic 1, 2, 3, 6, 8, 9, 10, 12, 21, 22, 47, 70, 75, 79-82, 85, 89, 98-102
Stress response 31
Surgery 2, 11, 12, 92

T

Thymic carcinoid 2, 6, 9-12
Transcription 17, 18, 20-23, 28-31, 37-40, 42, 43, 45-47, 52, 53, 59-61, 70-73, 75, 81, 87, 88, 91, 106, 111, 113
Transcriptional regulation 21, 45, 52, 54, 59, 87, 106
Transformation 23, 37, 40, 51, 54, 55, 110
Transforming growth factor-beta 69, 71-75, 79, 83-85
Trithorax 31, 51
Tumorigenesis 1, 21-23, 38, 47, 59, 69-72, 75, 79, 81, 82, 85, 87, 88, 97, 98, 102, 109, 112
Tumor suppressor 1, 2, 4, 5, 17, 20, 22, 27, 30, 37, 38, 40, 43, 44, 47, 53, 59, 60, 69-71, 74, 75, 80-82, 85, 87, 88, 91, 99, 105, 106

V

VIPoma 3, 7, 8